NO MORE MATH FACT FRENZY

DEAR READERS,

Much like the diet phenomenon *Eat This, Not That*, this series aims to replace some existing practices with approaches that are more effective—healthier, if you will—for our students. We hope to draw attention to practices that have little support in research or professional wisdom, and offer alternatives that have greater support. Each text is collaboratively written by authors representing research and practice. Section 1 offers practitioners' perspectives on a practice in need of replacing, and helps us understand the challenges, temptations, and misunderstandings that have led us to this ineffective approach. Section 2 provides a researcher's perspective on the lack of research to support the ineffective practice(s), and reviews research supporting better approaches. In Section 3, the authors representing practitioners' perspectives give detailed descriptions of how to implement these better practices. By the end of each book, you will understand both what not to do, and what to do, to improve student learning.

It takes courage to question one's own practice—to shift away from what you may have seen throughout your years in education and toward something new that you may have seen few if any colleagues use. We applaud you for demonstrating that courage and wish you the very best in your journey from this to that.

Best wishes,

—*Nell K. Duke and M. Colleen Cruz, series editors*

No More
Math Fact Frenzy

LINDA RUIZ DAVENPORT

CONNIE S. HENRY

DOUGLAS H. CLEMENTS

JULIE SARAMA

HEINEMANN
Portsmouth, NH

Heinemann
361 Hanover Street
Portsmouth, NH 03801–3912
www.heinemann.com

Offices and agents throughout the world

The authors and publisher wish to thank those who have generously given permission to reprint borrowed material:

"X-Ray Vision Game" from *Building Blocks Pre-K* by McGraw Hill. Copyright © 2013 by McGraw-Hill. Reprinted by permission of McGraw-Hill.

"Hitting a Target Number" from the Illustrative Mathematics website. Copyright © by Illustrative Mathematics. Reprinted by permission of Illustrative Mathematics.

"The Product Game" from *Illuminations* by the National Council of Teachers of Mathematics. Adapted with permission by the National Council of Teachers of Mathematics. All rights reserved.

Library of Congress Cataloging-in-Publication Data

Names: Davenport, Linda Ruiz, author. | Henry, Connie S., author. | Clements, Douglas H., author. | Sarama, Julie, author.
Title: No more math fact frenzy / Linda Ruiz Davenport, Connie S. Henry, Douglas H. Clements, Julie Sarama.
Description: Portsmouth, NH : Heinemann, [2019] | Series: Not this but that | Includes bibliographical references.
Identifiers: LCCN 2018051112 | ISBN 9780325107325
Subjects: LCSH: Mathematics—Study and teaching.
Classification: LCC QA135.6 .N64 2019 | DDC 510.71—dc23

LC record available at https://lccn.loc.gov/2018051112

Series Editors: Nell K. Duke *and* M. Colleen Cruz
Acquisitions Editor: Katherine Bryant
Production Editor: Sean Moreau
Cover Design: Monica Ann Crigler
Interior Design: Suzanne Heiser
Typesetter: Valerie Levy, Drawing Board Studios
Manufacturing: Val Cooper

Printed in the United States of America on acid-free paper
23   22   21   20   19   RWP   2   3   4   5   6

CONTENTS

This book, the fourteenth in the Not This, But That series, marks the first with coeditor M. Colleen Cruz. Coeditor Ellin Oliver Keene, who has been with the series since its inception, is moving on to other projects. It has been a privilege to work with Ellin. Professionally, I will be ever in awe of her thought leadership. Personally, I am so grateful to call her a dear friend.

I also want to take the opportunity to welcome M. Colleen Cruz as a coeditor. How lucky I am that Colleen Cruz—*the* Colleen Cruz—has agreed to take on this role. It has been a pleasure already to engage with her on this series, and I look forward to our continued collaboration.

—*Nell K. Duke, series coeditor*

INTRODUCTION

Nell K. Duke

I completely understand how we've ended up in a math fact frenzy. It seems as simple as $1 + 1$ that if we want kids to be able to retrieve math facts automatically, we should have them practice them over and over and over again as rapidly as possible. Plus, we've learned lots of ways to make that process seem fun, for example, in Around the World, Slam 10, and Math Facts Race.

The problem is that just being able to quickly spout out math facts does little for children if they don't have important underlying understandings. Without these understandings, many children will find it difficult to learn math facts in the first place. And what about those fun math fact competitions? When we look closely, we often find that those are fun for children who already know their math facts and far less fun for those who don't.

You want children to have more than math facts; you want them to be mathematical thinkers, to be aware of patterns, and to have tools to explore quantity, structure, and space. You want out of the frenzy and in to the calm of well-planned, well-executed—and, yes, fun—math fact education. To support you in that transition, the author team before you has expertise to the power of four. Linda Ruiz Davenport and Connie S. Henry are highly experienced practitioners who have helped countless teachers improve their practice. Douglas H. Clements and Julie Sarama are among the most respected mathematics education researchers in the United States, and their insights have been reshaping early math teaching.

Section 1 and the first portions of Section 2 identify the misconceptions and limitations of traditional approaches to developing math fact fluency. In the remainder of Section 2, you'll find lots of information pertaining to how children develop math fact fluency and how we can best support them in that process. In Section 3, you'll find models that help you envision your own better practice regarding math fact fluency along with nitty-gritty details about teaching math facts, including how to ask children specific questions to assess and prompt their thinking, how to make effective use of timelines, and what to do before, during, and after particular math games to make them educative and, yes, fun.

Like many of our Not This, But That titles, this book is likely to lead you to do things quite differently than the teacher down the hallway, the school down the street, or the district down the highway. Be confident that these authors don't take the call to diverge from traditional practice lightly. They write with a large body of research and experience behind them. I encourage you to subtract from your existing practices those that the authors do not recommend, add to your practices those they suggest, and watch the effectiveness of your math fact instruction multiply.

NOT THIS

Worksheets and Drill ≠ Math Fact Fluency

LINDA RUIZ DAVENPORT AND CONNIE S. HENRY

Remember back when we learned basic addition, subtraction, multiplication, and division facts in elementary school? We would have our pencils sharpened for our worksheets or our weekly timed test. There would be a page of fill-in-the-blank answers set up in neat rows and columns. We were told we would have three minutes to write down the answer to each math fact, and then the teacher would say "Begin!" and start the timer. We might have felt a sense of panic when our brains froze and we could not bring up the answers that only a few minutes earlier we knew.

Most of us really wanted to learn our math facts. Many of us practiced at home with flash cards, perhaps making our own in our neatest handwriting and grouping them by each number ($2 \times 1, 2 \times 2, 2 \times 3$). We recited them out loud to ourselves or with a family member, trying to memorize each one, only to be frustrated by facts that included larger digits, especially those 7s! We wanted to succeed on those timed tests, but somehow we never managed to do our best, as hard as we worked to prepare for them. How many students in math classrooms today have those very same experiences and those very same feelings?

What does it really take to help students learn their math facts in ways that allow them to access and use these facts fluently and flexibly to solve rich and challenging math problems? Are there strategies we could be using to help students learn their math facts more successfully, and with less stress and anxiety? In Section 2 we will hear more about what the research tells us about how students build their knowledge of math facts, and in Section 3 we will explore the kinds of classroom activities that help build that knowledge base, but for now, let's take a look at what we often see in elementary classrooms.

What We Often Encounter

Below, we describe a few scenarios that capture what we encounter as we visit classrooms where teachers, with the best of intentions, are attempting to address number fact automaticity. As you read through these scenarios, we invite you to reflect on why these practices, and others like them, might be problematic but at the same time seem to endure in so many classrooms.

Counting by Ones—Quickly!

This kindergarten classroom is colorful, welcoming, and full of energy. Students are working in centers and appear to be very focused on the tasks at hand. At one center, each student has a worksheet consisting of one hundred single-digit addition problems, all lined up vertically. At another center, students have a worksheet with one hundred single-digit subtraction problems, also lined up vertically. In both centers, students are quickly grabbing connecting cubes from a bin, quickly counting out the number of cubes needed to start the problem, then quickly adding more or taking some away, and then quickly counting the total—all out loud. For example, one student solves 9 + 3 by quickly counting out 9 cubes, then quickly counting out 3 more cubes, and then quickly counting all 12 cubes in front of her. She then very proudly and quickly writes the total on her worksheet.

My visit to this kindergarten classroom took place in the spring, and I was quite impressed with how quickly students, by this time of the year, were able to count collections of up to 20 objects and even beyond. I was also impressed with how they all seemed to understand that addition meant putting collections together and subtraction meant taking collections away.

But I couldn't help wondering how all this counting by ones was helping to build a foundation for learning number facts. True, students were very quickly getting right answers to these addition and subtraction problems. But what were they learning? Were they seeing any connections as they went from one addition problem to another or one subtraction problem to another? Did working so quickly get in the way of exploring these relationships? I also found myself wondering why they were doing all this work with addition and subtraction on separate worksheets, losing an opportunity to think about connections between addition and subtraction facts, or even to use addition facts to figure out subtraction facts.

Another question I asked myself was why they were doing number facts all the way up to totals of 20. It's true that you can use counting by ones to find those larger sums, just as you can use counting to find sums within 5 or 10. But why move into those larger numbers? These students did not seem to need to practice their counting, as they all seemed to know their counting sequences perfectly, so what were they learning about these larger combinations?

Finally, I also wondered if the vertical notation, in conjunction with the speed, was also preventing the students from making connections. As we learn in Section 2, the format of arithmetic fact practice matters. Horizontal number sentences that include an equal sign help students notice number relationships and build number fact fluency. Students who have opportunities to work with math facts expressed as *number sentences* (or equations, such as $3 + 4 = 7$) can begin to see relationships between each side of the equation, almost as if it were a balance scale. Adding something to one side of the equal sign has implications for what needs to change on the other side of the equal sign so everything can stay balanced. Seeing $2 + 3$ as "the same as" 5 can help us think about what might happen if we have 5 and then we take 2 or 3 away. It also invites a certain kind of playfulness: for instance, if $2 + 3$ is one way to make 5, what are other ways I could make 5?

But here, with so many problems on a single worksheet that were to be solved quickly, and with a vertical notation that reinforced the idea that each math fact was its own independent problem, students quickly counted by ones with cubes to solve each and every math fact as quickly as possible, with little conversation about what they were doing or what they were noticing. Had I asked any student to give me the answer to a math fact, I am confident they would have reached into the bowl for counters, counted by ones, and given me a correct answer. It looked like the goal was simply to finish the task using their cubes.

I could see that this lesson might look like a successful one in the moment, given the level of student engagement and the consistency with which students were getting correct answers. An observer in this classroom might think that students were well on their way to learning their addition and subtraction facts. But I left the classroom with lots of questions to consider about what might come next for these eager learners in terms of math fact fluency. For instance, did these students really need to count by ones with cubes to find each and every one of those math facts? Might some have been ready for slightly

> An observer in this classroom might think that students were well on their way to learning their addition and subtraction facts. But I left the classroom with lots of questions to consider about what might come next for these eager learners in terms of math fact fluency.

more efficient strategies, such as using their fingers to find facts like 2 + 3 or 3 + 2, or adding on to find 5 + 2 by starting with 5 and then adding on to get 6 and 7? Were students being held back by how the task had been presented?

How *do* students begin to move from counting by ones with counters or cubes to building the kind of number sense they need to learn their math facts, not just for addition and subtraction, but eventually for multiplication and division as well? How far can quickly counting by ones to find every math fact take you? (Section 2 aims to answer these questions, as well as others we raise in this section.)

Worksheets Once a Week

In a first-grade classroom, a teacher asks all the students to be silent as she passes out a traditional worksheet with thirty or more addition and subtraction number fact problems, all set up with vertical notation. The students know the drill and get to work right away, perhaps hiding their fingers under their desks as they try to figure out the different sums or differences. Their faces are mostly earnest, though several seem to have a worried look. After some time, the teacher collects the worksheets and students go off to lunch and recess, looking as if this break was perhaps a welcome respite from math. The teacher puts away the worksheets to score later. She plans to pass back the scored worksheets tomorrow so students can practice any facts they did not answer correctly using their flash cards.

There are indeed lots of addition and subtraction facts: 110 addition facts for numbers 0 to 10 and another 110 subtraction facts. As I watched, it was clear that most students knew some facts from memory but used their fingers to help them solve those they did not yet know (for instance, 4 + 1 = 5, 5 + 1 = 6, or 5 + 5 = 10). But I found myself wondering how these students would come to know all these facts for both addition and subtraction. Were these first-grade students able to make any connections from one math fact to another? Could they use the facts they knew to figure out the facts they were still learning? Or did they think their goal was to memorize each fact independently? Had anyone ever suggested that they use what they knew to figure out what they didn't yet know?

I also wondered what would happen if students felt they could use their fingers more openly as they worked on these math facts. For instance, using fingers to find that 3 + 4 is the same as 7 can also help you see that 4 + 3 is also the same as 7, just by switching hands. And by changing the 3 into a 4, you can see that 4 + 4 is the same as 8. As we learn in Section 2, fingers can serve as manipulatives that support student thinking—they are tools that you always carry with you. There is also something playful about using fingers, and I couldn't help wondering if that playfulness could also contribute to the spontaneous generation of several related math facts, just for fun! We also know that using fingers can help students learn to work with numbers and operations at a more symbolic level.

Finally, I wondered about the goal of these weekly worksheets that provided practice with math facts. Was the teacher hoping that by using these worksheets on a weekly basis, students would become more motivated to learn and remember their facts? Perhaps they would be more likely to practice with flash cards at home or during extra time between classroom activities? But how well would they remember all these facts over the long term? And what about students who just weren't making progress? Would they give up? Would they think that mathematics is a boring subject that requires lots of memorization? It was hard to think about students facing failure at this young age and what this might mean for how they saw themselves as mathematics learners.

> **It was hard to think about students facing failure at this young age and what this might mean for how they saw themselves as mathematics learners.**

Timed Tests: Knowing Math Facts Quickly

This third-grade teacher is starting math time with a timed test of multiplication number facts. The teacher distributes a problem sheet to each student, placing it face down so they cannot see the problems. She sets a timer for three minutes. "Ready, set, go!" she says as she starts the timer. Students quickly turn over their sheets and begin working, looking like they are concentrating hard as they write their answers to each problem, working as quickly as they can. Some students are further along than others. A few stop to erase an answer and replace it with what they think is a better choice. The alarm goes off. The teacher tells them to put their pencils down. There are a few sighs from students who didn't quite make it through the entire sheet. The teacher collects the sheets and lets them know she will be updating the wall chart showing how well each student was doing, adding a gold star for students getting all the problems correct. I could see that a few students had already earned their stars but some were still struggling to get there.

It looked like these timed tests were a classroom routine that students knew well. I could tell that some students had learned many of their multiplication facts and seemed to look forward to this opportunity to show off what they knew. But I could also see that many students looked uncomfortably stressed; they left some problems unanswered and some students completed only the first couple of rows, perhaps because they thought it best to solve each problem in the order given and had gotten stuck. It was clear that students could tell who knew their facts and who had not yet learned them, particularly with the wall chart showing everybody's progress.

I asked the teacher about these timed tests, and she responded that they were well worth the three minutes they took out of math class, given the importance of number fact fluency. She believed these timed tests communicated the importance of knowing math facts quickly, and because she gave them often, students had opportunities to improve so they could all eventually earn gold stars. I could see how this made sense, but how did improvement happen? I remembered back to my own experience rotely memorizing number facts, going through the stack of flash cards with my father, and finally getting all the multiplication facts correct. I also remember how quickly I forgot them! What was 7×8? 6×9? If I didn't keep practicing, some of these quickly faded from memory, coming back to me only after more work with those flash cards. But what about the students who just couldn't remember those facts from day to day, not to mention week to week?

> But is rotely memorizing facts really the key if what we've memorized seems to disappear from memory so quickly?

I could see how these timed tests, along with the wall chart showing progress, might create incentives for students to know their facts with automaticity. But is rotely memorizing facts really the key if what we've memorized seems to disappear from memory so quickly? What helps us remember these facts? For some reason, some facts seemed easier to hold on to, like the 5 times table, where I could skip count by fives, or the 4 times table, where I could use what I knew about the 2 times table to help me think about the 4 times facts. In Section 2, we have opportunities to think about the important role of reasoning strategies in learning all our math facts, but unfortunately many of these reasoning strategies remain invisible to students.

I also kept thinking about the stressed looks on so many of these students' faces. As we'll see in Section 2, research suggests that students whose teachers used timed tests knew fewer math facts than students whose teachers did *not* use timed tests. Does speed prevent students from taking the time to use their reasoning strategies? Does speed create a stressful situation that makes students tend to forget what they thought they knew? How many of us perform at our best when we are being timed, especially if we are still just learning? It seems like there is lots to think about with regard to timed tests of number facts.

Writing Down Our Math Facts to Help Us Remember Them

In this fifth-grade classroom, all the students are working independently on their multiplication facts. The teacher has given them a template with five columns on the front and five columns on the back, and their assignment is to list every fact, in order, from 1 to 10 in each column. As I watch, I can see students listing their facts—3 × 1 = 3, 3 × 2 = 6, 3 × 3 = 9—working down each column. The strategy most students seem to be using is skip counting, and I can see that sometimes they use their fingers to help them skip count as they get to the 6s, 7s, 8s, and 9s. Along the way I also see a few mistakes which throw off the multiplication facts that follow, but students don't seem to notice. They seem more focused on completing the task than on getting each fact correct. I ask a student if she can tell me what 5 times 7 equals. She smiles as she consults the "5" column to tell me it equals 35. I get the impression that the teacher asks students to create these tables of facts on a regular basis.

I wondered, did this teacher believe that by listing number facts in order, over and over again, students would eventually come to know each fact? Or did this teacher think that by making these lists, students would begin to see some patterns in their times tables that could help them learn their facts? This might have been true for some students, but others seemed to be counting by ones as they skip counted, especially once they got into higher numbers, sometimes with the help of busy fingers. The fact that students were using so much skip counting to complete their tables suggested to me that they were likely re-creating their number facts each time they made their tables. I wondered if a conversation about the patterns they were noticing in their tables might have been helpful.

As I thought about what I was seeing, I felt encouraged by the fact that students seemed to understand the idea of multiplication—they understood that to get from 3 × 6 to 4 × 6, you needed 6 more, or another group of 6. They also could easily find a fact they needed by looking in the right row in the right column, which showed they knew something about how times tables work. But this wasn't the same as knowing any single multiplication fact as it was needed. In Section 3, we'll see some learning experiences that can help students really know their number facts without having to re-create the table.

I was also concerned that students were often counting on by ones to get to the next multiple as they skip counted. This suggested that there was still some work for many of them to do with their addition facts. How can we help students learn their multiplication facts if they are still working on their addition facts? How far back do we need to go to help these students? Relationships among operations turn out to be key as students learn their math facts, and this includes relationships between addition and multiplication, subtraction and division, addition and subtraction, and multiplication and division. In Section 2, we'll learn more about the importance of these rich connections as students learn *and are able to remember* their math facts.

Number Games and Math Facts

In one third-grade classroom, students are sitting with partners in various parts of the room, playing number games. I know that such games are often considered a way to build number fact fluency, so I join a pair of students to see what they are doing and what number facts they are learning. One student rolls two number dice and gets a "3" and a "4," meaning he can move his piece 7 spaces along the path to the rocket ship. I hope to see this student add the 3 and 4 to get 7 and then move his piece 7 spaces, but this is not what happens. Instead, he moves his piece 3 spaces and then 4 more. What is this student having an opportunity to practice with this number game? I join another pair of students playing the game, this time with dotted dice. One student rolls a "5" and a "3." He then proceeds to count each dot—"1, 2, 3, 4, 5, 6, 7, 8"—and moves his piece 8 spaces. "Did you really need to do all that counting?" I asked. "No," he admitted. "I know 5 plus 3 is 8."

Many elementary math curriculum programs include number games that are intended to be a tool for helping students learn their math facts, and most elementary classrooms seem to have a collection of math number games on their shelves. These are intended to provide students with practice learning their math facts, often with a focus on addition or multiplication, depending on the game. Teachers typically introduce these games to students in the "rug area" of the classroom and then send them off to work in pairs, making sure they all understand the rules and know the expectations for working together as partners.

I knew from watching this teacher introduce the game that students were to get their game board pieces from home base to the rocket ship by rolling two dice, adding the two numbers together, and then moving their pieces that many places along the designated path. While one student added the numbers on the dice, the other was responsible for checking the sum and making sure the piece was moved exactly that many places. Some dice were dotted and some had numbers, and in some cases students had one of each, depending on which dice they drew out of the bin. I had the sense that this game was similar to many others these students had learned to play, and their focus seemed to be on getting their pieces to the final destination, rather than on practicing their math facts.

Students were certainly engaged and seemed to be having a good time together. But I also noticed that students were using quite a range of strategies to add their two numbers together. Some students using the dotted dice were counting by ones to get their total. Some students using a combination of a number die and a dotted die seemed to be adding on, sometimes using their fingers. Some students using number dice used their fingers, but more often they just moved the number of spaces indicated by the first die and then moved the number of spaces indicated by the second die. It was interesting for me to see that the strategies they used to add their two numbers seemed to depend on the kind of dice they had. This raised questions for me about how the tools students have inform the strategies they use when practicing their math facts. How do teachers decide which are the best tools to support students as they learn their math facts? Or do they think of dice as tools that are pretty much the same regardless of how numbers are represented on them? Some of these questions about choosing tools to support fluency in math facts are discussed in Sections 2 and 3, where you'll also find some advice about what different tools offer and how they might be used during classroom activities.

As the teacher circulated, there seemed to be little conversation about the strategies students were using and the extent to which they were drawing on their known number facts. But what incentives were there to draw on any known number facts, especially if their known math facts were not yet solidly held in memory? All of this made me wonder about exactly what students were practicing while they played this game and the extent to which it helped students learn their math facts with greater fluency. We will hear more about how to use games *effectively* to support math fact fluency in Sections 2 and 3.

Using "Fun" Worksheets and Puzzles

During one of my school visits, I stop by the copy room, since here I often get some insight into what resources teachers are using to teach math apart from what comes with the district's adopted math curriculum materials. I notice stacks of worksheets that did not copy correctly in the recycling bin. This worksheet is titled "Secret Multiplication Word Puzzle Valentine's Day—Sixes" and contains twenty multiplication problems, written vertically, all including a factor of 6. After students complete the twenty problems, they use a key at the bottom of the worksheet to convert products to letters that then spell out a holiday message.

Math fact worksheets seem to be everywhere, in one form or another. I often find them in copy rooms but I also see plenty of them in teachers' classrooms, organized into stacks at the side or back of the room. When I ask teachers about these worksheets, they often reassure me that these do not replace the math that is part of their lesson, but are designed to provide students with "extra practice" once they finish their work. An added bonus is the fact that many of these worksheets do not even need to be corrected, since checking the decoded message (or whatever else is built into the worksheet as the "reward") can tell both students and teachers whether success has been achieved.

I've noticed that students often complete these worksheets easily, partly because they might know some facts like 1 × 6 and 2 × 6 and partly because they can figure out the rest by counting, especially if they skip around the worksheet to do the facts in order, or if they do some skip counting on the side. I've also noticed that students often complete these somewhat leisurely; they are in no big hurry to get them finished, even with a message or some other "reward" for getting them all correct.

But what are students learning by completing these worksheets? Teachers tell me the extra practice provided by these worksheets helps support automaticity with math facts. I imagine that administrators and parents would also agree. But I would guess that if I asked students, "What's 4 × 6?" they would need to look at their worksheet to find the answer. Just writing down the answers to these multiplication problems doesn't seem to affect what they know or how quickly they know it, especially since they can work backwards from the message to figure out the fact they need. What might be a more meaningful alternative to these holiday worksheets?

Some Final Questions

In these visits, I can see that teachers value fluency with number facts, and I can see they are committed to helping students learn them, though I wonder how teachers are making sense of the standards indicating that students should learn these facts "from memory." But I also find myself thinking of all the middle school and high school math classrooms that I visit where I see so many students who still, as young adults, struggle with their math fact knowledge. I see some teachers using worksheets even with these middle school and high school students, but more often I see teachers who have given up on helping students learn their number facts, giving them multiplication tables to tape to their desk that they can consult as needed, or handing them calculators. But why is achieving number fact fluency so difficult for so many of our students? Why is it surrounded with so much stress? Why do so many students come to think of themselves as not good in math because they cannot get these number facts memorized?

It feels like something about the way we approach number fact knowledge is just not working for our students. Writing our facts over and over again, whether this work is a timed test or "dressed up" to seem like fun, doesn't really seem to help. Activities with "cute" contexts for finding number facts are often just an opportunity for students to keep using what they already know, even if it's counting by ones, because there is little incentive to move into less comfortable strategies that are new to them. So how do students learn their number facts? Why don't these kinds of approaches work? What *does* work? What does the research tell us?

WHY NOT? WHAT WORKS?

Authentic Fluency, Not Rote Memorization

DOUGLAS H. CLEMENTS AND JULIE SARAMA

Science is facts; just as houses are made of stones, so is science made of facts; but a pile of stones is not a house and a collection of facts is not necessarily science.

—Jules Henri Poincaré (1905, 141)

Anyone who wishes to be mathematically proficient needs to "just know" simple arithmetic— 4 + 9 = 13, for example, or 3 × 7 = 21. Students who do not learn such facts tend to have difficulties learning more advanced mathematics.

The reason for this is not that children cannot solve, for example, a multidigit arithmetic problem without knowing these basic facts, but rather that it is far more difficult to do so. When students achieve "automaticity"—easy access to these facts—they free up mental resources that they can use to think about other aspects of the math, such as place value, complexities of renaming, and, of course, the problem they were solving in the first place. In general, they can give more attention to more complex matters (e.g., looking for patterns). Without automaticity, students have a hard time learning multidigit calculations—both written and mental. They also face a roadblock in learning multiplication and division and then in understanding and later calculating with fractions. Thus, early lack of automaticity starts a cascade of frustration and difficulty, ending with a failure in algebra (Baroody, Bajwa, and Eiland 2009; National Mathematics Advisory Panel 2008).

As we saw in Section 1, many teachers assume that students must "just memorize" their basic facts by rote. To them, that implies timed practices, worksheets, and flash cards. If you have to memorize facts, you need to practice each one separately, again and again, quickly. They often believe, "That's how we learned them, and that worked well."

The problem is, worldwide research shows that the way most people in the United States think about arithmetic facts and children's learning of them, and the language they use, *may harm more than help.* To see why, we have to step back and rethink each of these assumptions.

Getting Our Facts Straight: Misconceptions That Harm Students

There are three primary *misconceptions* embedded in these assumptions:

- Facts are single, disconnected items that must be learned separately.
- "Learning your facts" means rote memorization.
- The best way to master facts and build fast recall is through lots of rote practice (flash cards, worksheets) and timed tests.

Misconception: Arithmetic Facts Are Disconnected Items

To many people, the term *fact* often means a piece of disconnected information. Some of us learned arithmetic at a time when many psychologists warned that $4 + 9 = 13$ must be studied separately from

13 − 4 = 9 or even from 9 + 4 = 13. But almost nothing in mathematics is disconnected. Think of multiplication and division. Many people don't actually know "division facts"—for 56 ÷ 7, they think, "What times 7 is 56? . . . 8!" Further, 2 × 8 is, of course, related to 8 + 8. (Note that we will focus on multiplication and division after reviewing the more numerous studies on addition and subtraction.) Mathematics is a system and a structure. These operations are related to each other.

Perhaps the word *fact* is just a term, and we should not be overly concerned about its implication of isolated pieces of information. However, it is important that educators understand and agree that students need to learn *related* facts, as the great mathematician Jules Henri Poincaré points out in the quote at the beginning of this section. And they *also* should learn about arithmetic properties, patterns, and relationships as they do so. Further, that knowledge, along with intuitive magnitude and other knowledge and skills, ideally is learned simultaneously and in an integrated fashion with knowledge of arithmetic facts. Thus, knowing an arithmetic fact well—that is, *fluently*—means far more than knowing a simple, isolated piece of information. For example, students notice that the sum of *n* and 1 is simply the number after *n* in the counting sequence, resulting in an integration of knowledge of facts with their well-practiced counting knowledge. Students who have learned all this have *adaptive expertise*—meaningful knowledge that can be flexibly applied to new, as well as familiar, tasks (Baroody and Dowker 2003).

Misconception: Learning = Memorization

Some readers might argue, "Then you have to memorize each fact and the connections between the facts. That's just *more* memorization." That is, indeed, a psychological viewpoint that is more than a hundred years old. In this view, each fact is stored separately, such as in the verbal statement "Four plus nine equals thirteen" (Baroody, Bajwa, and Eiland 2009). Furthermore, counting to figure out a fact is viewed as an immature strategy that must be stamped out. Instead, "memorizing the math facts" has always and continues to be a central focus of the mathematics curricula and textbooks, with pages and pages of worksheets asking students to memorize the answers (Fuson 2003).

In contrast, research suggests that producing basic facts is *not* just a simple process of "rotely memorizing" each fact and then mentally "looking each up" in our memories. Retrieval of arithmetic facts is part of a complex process. We therefore think about retrieval as a rapid process of gaining access to a result using automatic mental processes. Unlike "just memorizing facts by rote," in arithmetic many brain systems work together. For example, systems that involve working memory, executive (metacognitive) control, and even spatial "mental number lines" support knowledge of arithmetic facts. Further, for subtraction calculations, both the region specializing in subtraction *and* the region specializing in addition are activated. So when children *really know* $8 - 3 = 5$, they also know that $3 + 5 = 8$, $8 - 5 = 3$, and so forth, and that all these facts are *related*.

But wait: Didn't most of *us* learn by rote memorization? Didn't that work? No, most of us, especially those who succeeded in school, learned far more than just the rote verbalization. We learned to make sense of the *quantities* and *relationships* among facts. Thus, we "know from memory"—but that is based on a deep network of understandings and skills. We call knowing from memory in this way *fluency*: accurate knowledge and concepts and strategies that promote adaptive expertise. *Students who achieve this kind of fluency can reconstruct facts, use them to solve new types of problems, and more.* We do believe that retrieval—as a *part* of the learning and teaching process—is a very good thing. This is especially so when it is based on relationships and not just on rote memorization—when "knowing from memory" is based on understanding.

Indeed, research has shone a light on the dark side of teaching memorization only, and it's not pretty. Let's take a look at one "natural" study that happened in California.

California used to have standards consistent with those of the National Council of Teachers of Mathematics (NCTM), with balanced consideration of concepts, skills, and problem solving. After pressure from conservative groups, the state accelerated addition and subtraction basic-facts memorization (Henry and Brown 2008). That is, children were supposed to know all the basic addition and subtraction facts by the end of first grade. Further, "knowing" those facts was restricted to memorization. State legislators passed laws that any textbooks purchased for California in 2008 had to teach children to memorize all the facts in first grade, with little guidance for second grade.

Textbooks and teachers were thus to directly teach memorization using practices such as timed tests and flash cards. All these requirements and recommendations may have stemmed from a misunderstanding of the educational practices of high-performing countries such as Korea, China, Taiwan, and Japan (Henry and Brown 2008).

How did that work out for the teachers and students? Not well. Only 7 percent demonstrated adequate progress. Even among students from the highest-performing schools, fewer than 11 percent made progress toward the memorization standard equivalent to their progress through the school year. Barely a fourth of the students demonstrated retrieval of 50 percent or more of addition and subtraction facts.

Misconception: Students Learn Through Rote Practice

In the California results, two instructional practices were *negatively* related to basic-facts retrieval:

- use of California state–approved textbooks demanding basic-facts retrieval in first grade; and
- use of timed tests.

Students whose teachers relied on the basic-facts-memorization textbooks knew *fewer* facts. Indeed, they achieved about a third as well on basic facts as those who relied less heavily on these textbooks. Students whose teachers followed the advice of these textbooks not only didn't retrieve the facts but relied most heavily on low-level counting skills.

Similarly, students whose teachers used timed tests knew *fewer* facts. It may be surprising to find that practicing exactly what you think is the outcome skill *works against learning that skill.*

This is all counterintuitive to many people. If you want students to memorize, *teach them to memorize.* Right? Wrong—at least in the limited sense of direct teaching of *rote* memorization—it doesn't work. At best, it develops only *routine expertise* (Baroody and Dowker 2003). At worst, it doesn't even do that well.

Other practices were neither helpful nor harmful. Flash card use didn't hurt, but didn't support students' learning either. Neither did extensive work on small sums. Presenting easier arithmetic problems far more frequently than harder problems isn't a good idea. The opposite is the practice in countries with higher mathematics achievement, such as East Asian countries (National Mathematics Advisory Panel 2008).

What happened? What does this study teach us? *Memorization without understanding, drill without developing concepts and strategies, are not effective ways to teach or learn arithmetic facts, much less the edifice that is mathematics.*

Even the *format* of arithmetic "fact" practice can interfere with present and future learning. Consider how reasonable these tasks seem:

> **Memorization without understanding, drill without developing concepts and strategies, are not effective ways to teach or learn arithmetic facts, much less the edifice that is mathematics.**

$$3 + 4 = \underline{\quad}$$
$$5 + 9 = \underline{\quad}$$
$$6 + 0 = \underline{\quad}$$
$$8 - 7 = \underline{\quad}$$
$$9 - 5 = \underline{\quad}$$
$$5 - 2 = \underline{\quad}$$

Let's say that these tasks are done *after* students develop the concepts and strategies of arithmetic. What harm could come from such traditional practice of addition and subtraction facts?

Once again, research is clear. The more children do such traditional practice exercises, the *lower* their scores on equivalence problems such as 2 + 6 + 3 + 4 + 6 = 3 + 4 + __. U.S. students as a whole get *worse* on such problems from seven to nine years of age. Shockingly, even *undergraduates* given such traditional arithmetic practice get *worse* on equivalence problems (McNeil 2008; McNeil, Fyfe, and Dunwiddie 2015; McNeil et al. 2011).

A steady diet of such types of tasks teaches students limited patterns of thinking. They learn unfortunate rules such as "The equal sign means compute and put in the answer."

What Works? Achieving Fluency with Arithmetic Facts

Alex is five years old. Her brother, Paul, is three. Alex bounds into the room her father is in and this conversation ensues:

> ALEX: When Paul is 6, I'll be 8; when Paul is 9, I'll be 11; when Paul is 12, I'll be 14 [she continues until Paul is 18 and she is 20].
>
> FATHER: My word! How on earth did you figure all that out?
>
> ALEX: It's easy. You just go "three-FOUR-five" [saying the "four" very loudly, and clapping hands at the same time, so that the result was very strongly rhythmical and had a soft-LOUD-soft, ABA-ABA pattern], you go "six-SEVEN [clap]-eight," you go "nine-TEN [clap!]-eleven" [she continues until Paul is 18 and she is 20]. (adapted from Davis 1984, 154)

Alex had figured out, using her knowledge of the world and of numbers, how to add 2 to any number up to 18. Is this small, but remarkable, scene a glimpse at an exceptional child? Or is it an indication of the potential all young children have to learn arithmetic? If it's the latter, how early could instruction start? How early should it start?

Learning arithmetic facts well takes years. Adults use retrieval (that is, quickly remembering a fact) in only about 80 percent of situations (with operands less than 10) (LeFevre, Sadeskey, and Bisanz 1996). But even such retrieval isn't simply accessing memorized information, because many people use covert strategies, quickly and unconsciously calculating an answer based on related facts or other reasoning strategies (Bisanz et al. 2005; Brownell 1928).

The Phases of Learning Facts: Addition and Subtraction

Meaningful and effective learning of number facts involves three phases: (1) building foundational concepts of number and arithmetic and learning to figure out simple facts with counting and visually based strategies, (2) learning *reasoning strategies* to determine facts more efficiently, and (3) achieving full fact fluency (Baroody, Bajwa, and Eiland 2009; Clements and Sarama 2014a; Sarama and Clements 2009). Let's consider each in turn. We'll discuss the research here and provide suggestions for practice in Section 3.

Building Foundations and Initial Strategies

Knowledge of arithmetic *concepts* forms an organizing framework for storing arithmetic facts (Canobi, Reeve, and Pattison 1998). Students with greater conceptual knowledge are more likely to use sophisticated strategies *and* retrieve facts accurately.

To build foundational concepts and skills, we may have to change some of our beliefs. Most preschool teachers do not believe arithmetic is appropriate, nor do they even think of it as a topic (Sarama 2002; Sarama and DiBiase 2004). In multiple countries, professionals in many educational roles vastly underestimate beginning students' abilities (Aubrey 1997; Van den Heuvel-Panhuizen 1990). And kindergarten teachers tend to teach mostly concepts and skills that their students already know (Engel, Claessens, and Finch 2013). Children can learn and do much more.

With more positive beliefs, early childhood teachers can do much to build strong foundations and initial arithmetic strategies. They can ensure that children can count with understanding and apply counting strategies to solve problems. As an example, consider the activity "How many now?" (Clements and Sarama 2007–2013), in which children count several objects. The objects are then hidden, one is added, and the children are asked, "How many now?" When children are able, warn the children, "Watch out! I'm going to add more than 1 sometimes!" and sometimes add 2, and eventually 3, to the group. If children seem to need assistance, have a puppet model the strategy; for example, "Hmmm, there's *4*, one more makes it 5, and one more makes it 6. 6, that's it!" Such activities help children *relate* counting ("the next number") to addition ($n + 1$). Instruction such as this has proven effective in building strong early number and arithmetic competencies (Clements and Sarama 2007).

In preparation for multiplication and division, children can make equal groups, as in making "fair shares." They can also learn skip counting, first by tens, twos, and fives. A simple example is counting all the fingers in a classroom by counting by tens.

Another research-based approach involves developing students' skill in *subitizing*—the rapid perception of how many in a group. Teachers first build *perceptual* subitizing of small groups (1 to 4 or 5). For example, they play Snapshots. Teachers hide a collection of 1 to 4 objects arranged in a line or other simple arrangement, then show it for two seconds or less, hide it again, and then ask children to respond verbally with the number name. After children are comfortable and competent, they play Snapshots with different arrangements, such as the arrangements of five pictured in Figure 2–1, to develop *conceptual* subitizing—recognizing *two* groups separately and mentally combining them quickly (Clements and Sarama 2007–2013). The goal is to encourage students to "see the addends and the sum as in 'two olives and two olives make four olives'" (Fuson 1992, 248).

A benefit of subitizing activities is that different arrangements suggest different views of a number—that is, different number facts. For example, look at the dot arrangements in Figure 2–1. What combinations does each suggest to you? Can you see any in more than one way?

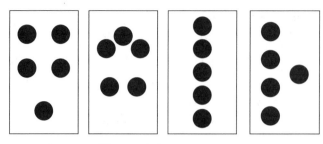

Figure 2–1 Dot patterns

Working with objects and story contexts (e.g., animals in 2 different pens) also helps children come to see all the different number facts for a given number (e.g., separating 5 objects into groups of 4 and 1 or 3 and 2). Research shows that preschoolers and kindergartners can learn these competencies and that their arithmetic improves as a result (Clements and Sarama 2008a; Clements et al. 2011; Clements et al. 2013). How well students subitize predicts knowledge of cardinality, knowledge of arithmetic, and general mathematics achievement well into the primary grades (Clements, Sarama, and MacDonald 2017).

When children can do this, they can be challenged to conceptually subitize a number of equal groups, such as 3 groups of 5 dots. This forms a visually based foundation for thinking about multiplication.

Students learn facts better if they invent, use, share, and explain *different* strategies.

Learning Reasoning Strategies

Effective teachers attend explicitly and directly to the important conceptual issues students are more likely to encounter. They help students develop important conceptual understandings, as well as procedural skills (Hiebert and Grouws 2007).

Students learn facts better if they invent, use, share, and explain *different strategies* (Baroody and Rosu 2004). Indeed, the *number of different strategies children use predicts their later learning* (Siegler 1995).

What are the best strategies? Research is clear that effective strategies include counting strategies (such as "doubles plus one" and counting on), conceptual subitizing, and break-apart-to-make-ten, or

BAMT (Baroody 1987; Baroody, Bajwa, and Eiland 2009; Henry and Brown 2008; Murata 2004; Murata and Fuson 2006). We will discuss each of these in Section 3.

Achieving Full Fact Fluency

To achieve full fact fluency, children must practice their arithmetic strategies. This is not drill but repeated encounters combining different ideas, and experiences for learning and internalizing them. Children might engage in whole-group (choral responding), individual-within-whole-group, and independent practice. In individual-within-whole-group practice, individual students answer but then ask the class, "Is it OK?" The class calls a response back. All this practice needs to emphasize conceptual links so students can synthesize knowledge for fluency and understanding. They can then begin distributed practice (practicing a variety of facts and strategies over time, rather than practicing one thing at a time for long periods). This is not rote learning or rote practice but a clear, high-quality use of skills and understandings to solve a variety of problems.

The Phases of Learning Facts: Multiplication and Division

Most research findings and implications regarding addition and subtraction facts apply to multiplication and division facts. Some strategies that students use are also similar, though they focus on equal groups. There are some strategies that are unique to multiplication and division facts, though, and we'll discuss those here.

Building Foundations and Initial Strategies

The developmental progression of multiplication strategies is similar to that for addition (e.g., Fuson 2003; Steffe 1994). At first, children can pass out objects to different people (or dolls, etc.), but they just "dump" the objects, without appreciation of equal numbers (Miller 1984). They begin with basic notions of "fairness" to build equal groups and to share objects equally among, first, two people, and later, larger numbers of people. Most do distributive sharing, distributing one at a time, but sometimes they still don't appreciate the numerical result—that is, they may make equal shares and yet do not explicitly recognize that if there are 7 in one share, there are 7 in the other share(s) (Bryant 1997; Miller 1984). Over time, they become more systematic, giving each person an object, checking that each has one, and repeating (Hunting and Davis 1991). Some students keep using these strategies until they are up to nine years of age (Miller 1984).

Eventually, students learn to solve small-number multiplication problems by grouping—modeling each of the groups (Carpenter et al. 2014). They can solve division or sharing problems with informal strategies, using concrete objects, now up to 20 objects and from 2 to 5 people. That is, some give 1 (or more) object(s) to each person, "dealing them out." Others do distributive counting or count out groups, such as giving 2 to each person again and again. Still others might count out equal groups, such as 5 for each person, then check if the sum equals the product, and adjust as needed.

Learning Reasoning Strategies

Next, students learn the inverse relationship between multiplication and division. They also learn to use count-based strategies, such as skip counting, to solve multiplication and measurement division problems. In measurement division, you know the total number, such as 28 candies, and how many you wish to give out to each person, such as 4 candies, and the question is, how many people will get a share? Children may solve 4×5 by skip counting by fives and using their fingers to keep track of the counts: 5, 10, 15, 20. Note that counting strategies are not used nearly as early for these operations as they are for addition and subtraction.

Eventually, children learn some facts, especially doubling ("times 2"). They then invent derived facts, such as solving 7×6 by knowing that five 7s is 35 (from previous skip counting or other experiences), so 7 more is 42. They learn to work with arrays—area models in rows and columns, extending their ability to use skip counting and multiplication in different contexts.

Then, ideally, children continue to use such reasoning strategies, using the commutative and associative properties (informally). For example, they might say "9 × 2 is nine 2s, but it's the same as two 9s, so 18." They also find and use patterns in the multiplication table, which promotes much easier and faster fluency with multiplication facts (Fuson 2003). Think of the 9s facts (9, 18, 27, 36, 45 . . .).

Achieving Full Fact Fluency

Achieving full fact fluency for multiplication and division is a long process, but it is similar to the process for addition and subtraction. Children practice their arithmetic strategies, which should link with conceptual understandings. They can then begin distributed practice, which again can involve some game-like drill, but also application of skills and understandings to solve a variety of multiplicative problems.

Research-Supported Instruction for Developing Strategies

Research also sheds light on a couple of issues regarding teaching reasoning strategies. Again, we'll highlight the research findings here, then dive into practical details in Section 3.

Invention First, More Efficient Strategies Later: Allowing children to invent their own strategies initially, rather than teaching all strategies directly, may have benefits, especially in children's development of mathematical concepts and problem solving skills (Kamii and Dominick 1998). Children's *making sense* of mathematical relations is key. A nine-month training experiment evaluated the effectiveness of a number sense curriculum designed to help at-risk preschoolers develop the prerequisite knowledge and discover the key mathematical relations underlying fluency with $n + 0$, $0 + n$, $n + 1$, and $1 + n$ facts (Baroody, Eiland, and Thompson 2009). Children's number sense improved and their knowledge of facts was therefore based on understanding.

Once children have had the opportunity to invent their own strategies and discuss different strategies with their peers, encouraging them to adopt more sophisticated, beneficial strategies may be possible with no harmful effects. The most effective approach is teaching children conceptually—emphasizing conceptual knowledge initially and in parallel with procedural knowledge (Rittle-Johnson and Alibali 1999).

Thoughtful Use of Manipulatives: Paradoxically, those who are best at solving problems with objects, fingers, or counting are least likely to use those less sophisticated strategies in the future, because they are confident in their answers and thus move toward accurate, fast retrieval or composition (Siegler 1993). Manipulatives can be necessary at certain stages of development. Preschoolers can learn nonverbal and counting strategies for addition and subtraction (Ashcraft 1982; Clements and Sarama 2007; Groen and Resnick 1977), but they may *need* concrete objects to give meaning to the task, the count words, and the ordinal meanings embedded in the situations. For the youngest children, use of physical objects related to the problem, compared to structured "math manipulatives," may support use of their informal knowledge to solve the arithmetic problems (Aubrey 1997). In certain contexts, older students too may need objects to count to create the numbers they need to solve the problem (Steffe and Cobb 1988).

Early use of manipulatives can help, not hurt. Teaching children to use their fingers as manipulatives in arithmetic accelerated children's single-digit addition and subtraction as much as a year over traditional methods in which children count objects or pictures (Fuson, Perry, and Kwon 1994). If teachers try to eliminate finger use too soon, children hide their fingers—which are then not as visually helpful—or adopt less useful and more error-prone methods (Fuson, Perry, and Kwon 1994; Siegler 1993). Further, the more advanced methods of finger counting are sufficiently efficient to allow multidigit and more complex computation and are not crutches that hold children back (Fuson, Perry, and Kwon 1994; see also Crollen and Noël 2015; Price 2001).

Moving Beyond Manipulatives: Once children have established successful strategies using manipulatives, they can often solve simple arithmetic tasks without them. Kindergartners given and not given manipulatives showed no significant differences in accuracy or in the discovery of arithmetic strategies (Grupe and Bray 1999). The similarities go on: children without manipulatives used their fingers on 30 percent of all trials, while children with manipulatives used them on 9 percent of the trials but used their fingers on 19 percent of the trials for a combined total of 28 percent. Finally, children stopped using such external aids approximately halfway through the twelve-week study:

As one student described it,

> I find it easier not to do it [simple addition] with my fingers because sometimes I get into a big muddle with them [and] I find it much harder to add up because I am not concentrating on the sum. I am concentrating on getting my fingers right . . . which takes a while. It can take longer to work out the sum than it does to work out the sum in my head. (Gray and Pitta 1997, 35)

By "in my head," Emily meant that she imagined dot arrays. If that's what she liked, why didn't she just use those images? Why did she use fingers? She explains:

> If we don't [use our fingers] the teacher is going to think, 'Why aren't they using their fingers . . . they are just sitting there thinking' . . . we are meant to be using our fingers because it is easier . . . which it is not. (Gray and Pitta 1997, 35)

In summary, students who learn to use different strategies in concert with meaningful recall are more likely to develop mathematical proficiency than those who have memorized the facts without developing their arithmetic strategies. One study compared two approaches to teaching addition. The traditional "atomistic" approach emphasized small numbers initially, addition as separate from subtraction, and *procedural* (not meaningful) use of physical manipulatives. The holistic approach moved from manipulative strategies to focus on counting strategies and the number system (Price 2001). Children worked with numbers of different sizes, patterns in number and arithmetic, the inverse relationship between addition and subtraction, and cardinal and ordinal representations. They applied counting strategies to solve problems, including skip counting by tens to 1,000. Children taught using the traditional approach made less progress than those taught using a holistic approach.

Building Fluency

Research has provided several guidelines for helping children achieve fluency with arithmetic facts. Recall that fluency combines accurate knowledge of concepts *and* strategies that promote *adaptive expertise*.

Provide Distributed Practice

To develop fluency, children need considerable practice, distributed across time (Ericsson, Krampe, and Tesch-Römer 1993). For example, rather than studying one fact for a minute, students should study it once, then study another fact, then return to the first one. Practice on facts is best done in short but frequent sessions. For long-term memory, a day or more should eventually separate these sessions (Clements and Sarama 2014b).

Use Practice for Speed Judiciously

Although timed tests are often done badly—remember the California study (Henry and Brown 2008)—practice for speed, done well, is useful and important. Tutoring with a small bit of speeded practice is more effective than tutoring with unspeeded practice (Fuchs et al.

2013). Combined with about twenty minutes of instruction on number knowledge and relationships, including emphasis on retrieval as well as efficient counting strategies for correcting any mistakes, five minutes of speeded practice leads to fluency and competencies in complex calculations (Fuchs et al. 2013). It is important to match instructional strategies with students' needs. Children who are accurate but slow may benefit

> Children who are accurate but slow may benefit from timed practice, but such timed practice may harm children who are struggling to be accurate.

from timed practice, but such timed practice may harm children who are struggling to be accurate (Codding et al. 2009). Once children are accurate and competent with strategies, game-like, self-motivated practice for speed is an excellent complement. Such practice should be short, frequent, stress-free, and fun, with students engaged in improving their own performance.

Provide Practice That Supports Strategy Use

Practice should not be "meaningless drill," but should occur in a context of making sense of the situation and the number relationships. Multiple strategies help build that number sense, and children who are strong in calculations know and use multiple strategies. This suggests that it is not just the arithmetic facts that should be automatic. Students should also be fluent with the related reasoning strategies. As an example, children might understand that problems such as "I have $8 but it costs $13 to buy a toy. How much more do I need?" and "What is 13 take away 8?" can both be solved by counting up from 8 to 13, keeping track of counts, or subtracting 13 from 8. Therefore, practice that includes many problems that can be solved with the same or similar strategy is particularly useful.

Even the *format* of arithmetic "fact" practice can interfere with present and future learning (McNeil 2008; McNeil et al. 2012; McNeil, Fyfe, and Dunwiddie 2015; McNeil et al. 2011). Simply providing practice that is better planned and presented can make all the difference. As an example, start without symbols so children think about the *quantities*; for example, ask, "What is 9 take away 4?" rather than writing:

$$9$$
$$\underline{-\ 4}$$

Then, after symbols are introduced, continue to use horizontal number sentences, leaving spaces around the equal sign, and avoid saying, "What's the answer?" Why? First, unlike the vertical form, which just indicates "write the answer," the horizontal form *does* have an equal sign. Second, because students often interpret the equal sign as indicating not equality or balance but "write the answer next," we replace "What's the answer?" with more helpful and interesting questions such as "What goes in the blank to make the sentence true?" Also, we present practice in different formats, which reduces reliance on limited operational patterns. Here are three:

1. *Switch the "sides" of the number sentence.*

$5 + 3 = 8$	$3 + 5 = 8$	$8 - 5 = 3$	$8 - 3 = 5$
$8 = 5 + 3$	$8 = 3 + 5$	$3 = 8 - 5$	$5 = 8 - 3$

2. *Continue to use words.* For example, ask, "What equals three plus four?"

3. *Group by equivalent values.* For example, it is not helpful to group by the "same addend" ($3 + 1, 3 + 2, 3 + 3$). Instead, consider the following group (McNeil et al. 2012):

   $$3 + 4 = \underline{\quad}$$
   $$5 + 2 = \underline{\quad}$$
   $$6 + 1 = \underline{\quad}$$

Such grouping helps students build knowledge of not only addition facts but also equivalence. As another example, teachers make "math mountain" cards such as those in Figure 2–2 (Fuson and Abrahamson 2009). Students cover any of the three numbers and show the other two to their partner, who tells what number is covered. Karen Fuson and Dor Abrahamson (2009) use other representations that show the part-part-whole relationship, such as dots and corresponding symbols (see Figure 2–2).

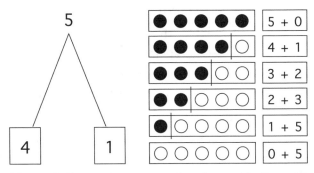

Figure 2–2 Multiple representations for combinations of 5

Similarly, for multiplication and division, feature different formats of multiplication sentences. Students should learn and use all the different symbols for division as well, including $24/8$, $24 \div 8$, and $\frac{24}{8}$.

Teachers can help build their students' fluency in multiplication and division by emphasizing the use of skip counting, reasoning strategies, and commutativity ($4 \times 7 = 7 \times 4$) as well as the search for patterns. Division facts may best be learned in relation to multiplication facts. For example, $48 \div 6 = \underline{\quad}$ can be solved by recalling $6 \times 8 = 48$.

Choose Digital Tools Carefully

One of the most common uses of computer-assisted instruction (CAI) is to provide practice, for example, in addition facts (Fuchs et al. 2006). Some reports say that the largest gains from the use of CAI have been in practicing mathematics for lower-primary-grade students (Fletcher-Flinn and Gravatt 1995), especially in compensatory education programs (Lavin and Sanders 1983; Ragosta, Holland, and Jamison 1981). About ten minutes per day proved sufficient time for significant gains; twenty minutes was even better. Another program showed good effects on fluency in arithmetic for a first grader who practiced for fifteen minutes three times per week for four months (Smith, Marchand-Martella, and Martella 2011). This CAI approach may be as or more cost-effective than traditional instruction (Fletcher, Hawley, and Piele 1990) and other instructional interventions, such as peer tutoring and reducing class size (Niemiec and Walberg 1987). This approach has been successful with students with different ability levels (Shin et al. 2012), with substantial gains reported for children from low-resource communities (Primavera, Wiederlight, and DiGiacomo 2001).

Technology practice can be especially helpful for children who have mathematical difficulties (MD) or mathematical learning disabilities (MLD) (Harskamp 2015). Of course, this should be introduced at the right point for the child's development and should be the right kind of practice. For example, "bare bones" practice, such as repeated, speed-based drill on arithmetic "facts," does not help children who are at the level of more immature counting strategies. Instead, research suggests using practice that helps them understand the concepts and learn the arithmetic fact rather than time-pressured drill (Hasselbring, Goin, and Bransford 1988). Also, practice that teaches fluency and cognitive strategies may be more effective, especially for boys (Carr et al. 2011).

This is an important point. In keeping with what we have learned, practice software that includes research-based strategies is more likely to be helpful (Fuchs et al. 2006; Powell, Fuchs, and Fuchs 2011). For example, the *Building Blocks* software not only provides the drill problems following these guidelines but also presents each group of facts based on the strategy that is most helpful in a particular type of solution. As a specific illustration, the software initially groups together all those facts that yield nicely to the break-apart-to-make-ten (BAMT) strategy. As another example, CAI has been found to be a feasible means of helping first graders with a risk factor discover the add-one rule (adding 1 is the same as "counting 1 more") by way of pattern detection (Baroody et al. 2015). The software might pose, "What number comes after 3 when we count?" and then immediately present a related addition question, "3 + 1 = ?" Also, an "add-zero" item and an addition item (with both addends greater than 1) serve as non-examples of the add-one rule to discourage overgeneralizing this rule. A similar technology program that combines fluency and cognitive strategy use helps second graders, especially boys, improve their arithmetic achievement (Carr et al. 2011).

In summary, technology-based, speed-oriented practice, especially with young children, should be used only when students are already accurate, confident, and strategic, and even then it should be used carefully and in moderation. Students' creativity may be harmed by a consistent diet of drill (Haugland 1992). Some students may be less motivated to perform academic work or less creative following a steady diet of only drill (Clements and Nastasi 1985; Haugland 1992; Sarama and Clements, forthcoming). There is also a possibility that children will be less motivated to perform academic work following drill (Sarama and Clements, forthcoming) and that drill on computers alone may not generalize as well as paper-and-pencil work (Duhon, House, and Stinnett 2012). In contrast, practice that encourages the development *and* use of strategies, that provides different contexts (supporting generalization), and that promotes problem solving may be more appropriate instead of, or in combination with, drill. (We'll look at criteria for choosing math apps and programs a bit more in Section 3.)

Properly chosen, computer games may also be effective (e.g., Ketamo and Kiili 2010). Second graders with an average of one hour of interaction with a technology game over a two-week period responded correctly to twice as many items on an addition facts speed test as did students in a control group (Kraus 1981). Even younger children benefit from a wide variety of technology-based games as well as nontechnological ones (Clements and Sarama 2008b). For example, in one simple game, young children place finger combinations on an iPad to play a game of recognizing and representing numbers before time runs out. Early pilot work with this novel interface is promising; it promotes children's use of their most accessible manipulative, their fingers, and provides a way to develop fact fluency based on meaningful actions (Barendregt et al. 2012).

A Short Warning About Addition and Multiplication: Watch for Interference

Taught well, addition and multiplication facts can support each other. But effective teachers know they can also interfere with each other. For example, when children begin learning multiplication facts, their addition performance declines (Fuson 2003). Especially confusing for students is the difference between addition and multiplication for facts involving 0 and 1. Talking about what adding 0 or 1 means, compared to multiplying by 0 or 1, before practicing the facts, can be useful for students.

Children with Mathematical Learning Difficulties and Disabilities

Many children show specific learning difficulties in mathematics. Two categories are often used (Berch and Mazzocco 2007). Children with *mathematical difficulties* (MD) struggle to learn math for any reason. Estimates range from 35 to 48 percent of the population. Those with a specific *mathematics learning disability* (MLD) are thought to have some form of memory or cognitive deficit that interferes with their ability to learn concepts and/or procedures in one or more domains of math (Geary 2004). They are, therefore, a small subset of all those with MD, with estimates of 4 to 10 percent of the population (Berch and Mazzocco 2007; Mazzocco and Myers 2003).

MD, MLD, and Arithmetic

One of the most consistent findings is that children with MLD have difficulty quickly retrieving basic arithmetic facts. This has been hypothesized to be due to an inability to store or retrieve facts, including disruptions in the retrieval process, and to impairments in visual-spatial representations. Difficulties in working memory and speed of processing have also been reported (Geary et al. 2007).

Others find that the retrieval times of children with MLD can be explained by the same factors that are thought to underlie performance limitations of normally achieving children (Hopkins and Lawson 2004). So they may *not* have an impaired working memory or a special kind of "retrieval deficit," but instead may have other difficulties—which are often simply a result of lack of appropriate instruction.

Still others have posited that impairments in understanding, learning, and remembering verbal material (Berch and Mazzocco 2007) may prevent these children from learning basic arithmetic combinations or facts. So, even by second grade, children with MLD may not understand all the counting principles and may have difficulty recognizing errors and keeping them in working memory long enough to fix them. They make more counting errors and persist in using developmentally earlier counting strategies. Indeed, they may continue to use immature, "backup" strategies with small variation and limited change throughout elementary school (Ostad 1998). The immature counting knowledge of MLD children and their poor skills at detecting counting errors may underlie their poor computational skills in addition (Geary, Bow-Thomas, and Yao 1992). Some only abandon finger counting by the end of the elementary grades, which is way too late; while, as we stated, early finger use is good, if it persists, it actually stalls development of fluency. Children with MLD often also have difficulty retrieving arithmetic facts, and although the other skills develop slowly, retrieval of facts does not improve for most children classified as having MLD (Geary, Brown, and Smaranayake 1991). These children may have difficulty ignoring irrelevant associations (e.g., for 5 + 4, they might answer "6" because it follows 5), as well as difficulty representing and manipulating information (Geary 2004). Still others may struggle with numerical ideas such as subitizing (Berch and Mazzocco 2007).

Accordingly, children who have MLD or MD likely have quite diverse learning needs (Dowker 2004; Gervasoni 2005; Gervasoni, Hadden, and Turkenburg 2007). These include support for gaining basic fact knowledge, carrying out arithmetic procedures, understanding and using arithmetic principles and estimating, acquiring other mathematical knowledge, and applying arithmetic in solving problems (Dowker 2005). Foundational abilities in subitizing, counting and counting strategies, simple arithmetic, and magnitude comparison are critical for young children with MLD (Aunola et al. 2004; Geary, Hoard, and Hamson 1999; Gersten, Jordan, and Flojo 2005; Holmes, Gathercole, and Dunning 2009).

Most delays in learning are due to a lack of opportunity to learn.

What all this means is this: educators must be extremely careful in labeling and working with children who show any signs of difficulty. Only a very small percentage of children have cognitive impairments; *most delays in learning are due to a lack of opportunity to learn* (Baroody 2011).

Moreover, putting the "blame" on the individual child not only is usually inaccurate but also lowers expectation for the very few who may have cognitive disabilities and prevents them from realizing their potential (Baroody 2011).

Regrettably, students identified as struggling in math are often given only drill on simple facts, rather than interesting work on concepts and problem solving. Special education teacher licensing programs often have few or no courses on mathematics or mathematics education, exacerbating this unfortunate situation. We can correct this by recognizing that students have different learning needs and thus a one-size-fits-all approach will not work for many, even most. Given lack of experience, some students may not notice number relationships as easily, so explicit strategy instruction may be necessary. This should be followed by "think-alouds" in which students use the strategy and explain their thinking (National Mathematics Advisory Panel 2008), as well as many opportunities to apply strategies to solving problems, and maybe even modify or invent strategies for themselves in future contexts.

Supporting Children with Identified Math Learning Challenges

All the recommendations in this book apply to most children with mathematics difficulties. Additionally, educators should enroll children with MD and possibly MLD in a research-based mathematical intervention as soon as possible. The children's progress in such an intervention can help identify children who may have been miseducated and mislabeled. Such an intervention would develop conceptual understanding, adaptive competence, and a productive disposition *along with* procedural fluency (National Research Council 2001). A guided investigative approach is useful in developing these simultaneously (Baroody 2011). Children who do not benefit substantially from these approaches, and who may have MD or MLD, should be provided specialized instruction—and not drill. Focus should go beyond arithmetic facts to include essential areas such as components of number and spatial sense.

For example, counting on fingers might be encouraged for a longer period of time for some students. Some children may have difficulty maintaining one-to-one correspondence when counting or matching. They may need to physically grasp and move objects, as grasping is an earlier-developing skill than pointing (Lerner 1997). Others may struggle with subitizing, counting objects in small sets one by one long after their peers are strategically subitizing these amounts. Emphasizing their ability to learn to subitize the smallest numbers, perhaps representing them on their fingers, may be helpful. (Children who have continued difficulty perceiving and distinguishing even small numbers are at risk for severe general mathematical difficulties [Dowker 2004].) Other children may have difficulty with magnitude comparisons (e.g., knowing which of two digits is larger) and in learning and using more sophisticated counting and arithmetic strategies (Gersten, Jordan, and Flojo 2005; Landerl, Bevan, and Butterworth 2004).

Again, this speaks to these children's diverse learning needs and educators' responsibility to use formative assessment and research-based learning trajectories (Clements and Sarama 2014a; Sarama and Clements 2009). Developing fact fluency should be done within a wider educational plan to develop number sense. Again, this is not different from the education of other children, but it is especially important for those struggling—and, sadly, is different from what these students usually experience. Number sense includes using different strategies for figuring out facts, using properties such as commutativity to reduce the number of facts to be learned, and generally looking for and using patterns and relations, such as $n + 0 = n$ (Baroody 2011).

The goal should be fact fluency. Strategies such as quickly thinking "$6 + 6 = 12$ to $6 + 7 = 13$" or "7×5 is seven 5s, or three 10s and $5 \dots 35$" may be the most meaningful route for some children. Practice should make *these strategies more efficient*, rather than quashing them with drill on isolated, unrelated facts. Most practice can be in the context of solving problems.

One more time: if these recommendations seem similar to those already discussed, you have interpreted us correctly. Extra patience, as well as more time for students to learn, is often required but will be well rewarded. This is *not* to say that no adaptations are needed. Planning supportive organizations, presentations, and uses of materials and adaptations of instruction; learning environments; and homework assignments can be useful and is often necessary (Shih, Speer, and Babbitt 2011).

Teach for Equity: An Asset-Based Approach

Finally, the educational research's most important implication for early childhood may be that we should try to prevent learning difficulties by providing high-quality mathematics education to all children. Equity must be complete equity, devoid of labeling, prejudice, and unequal access to opportunities to learn (Bishop and Forgasz 2007). Students from low-resource community and minority groups, and girls, often get less problem-solving experience and more routine drill than others do (Boaler 2002). Instead, as Sylvia Celedòn-Pattichis and her colleagues point out, we need to take an asset-based approach:

> Asset-based approaches to mathematics education are a conscious way to move away from deficit perspectives that view students, parents, and communities as lacking in different aspects that enable them to be ready for schooling. . . . An asset-based approach is grounded in the belief that students', families', and communities' ways of knowing, including their language and culture, serve as intellectual resources and contribute greatly to the teaching and learning of high-quality mathematics. . . . This approach draws from funds-of-knowledge work in which researchers and teachers learn with and from students, parents, and communities [in which] different ways of doing mathematics are acknowledged and honored. (Celedòn-Pattichis et al. 2018, 375).

We need to build on students' community and family "funds of knowledge" (Moll et al. 1992) *and* use research-based instructional approaches using that knowledge. All children can learn fact fluency if they are taught in a way that builds on what they know and what they have experienced.

Final Words

An important goal of early mathematics is students' flexible, fluent, and accurate knowledge of arithmetic facts. Learning these facts is not about rote memorization. Seeing and using patterns, and building relationships, can free children's cognitive resources to be used in other tasks. Children generalize the patterns they learn and apply them to facts that were not studied (Baroody and Tiilikainen 2003). Number fact instruction that focuses on encouraging children to look for patterns and relations can generalize to problem-solving situations and can free attention and effort for other tasks. It develops students' *adaptive expertise*. As we saw earlier, this kind of instruction involves three phases: (1) building foundational concepts of number and arithmetic and learning to figure out simple facts with counting and visually based strategies, (2) learning *reasoning strategies* to determine facts more efficiently, and (3) achieving full fact fluency. We take a detailed look at instructional practices in Section 3.

SECTION 3

BUT THAT

Strategies for Building
Authentic Fact Fluency

LINDA RUIZ DAVENPORT AND CONNIE S. HENRY

How can we help all our students achieve fluency with basic number combinations? What approaches, strategies, activities, and supports can we use? In this section, we'll draw on the research summarized in Section 2 and offer strategies to put those ideas into practice. First, we'll examine the teacher's role and summarize some general ideas and suggestions that apply to all the operations. Then, for each pair of operations (addition/subtraction and multiplication/division), we'll set the stage and identify key models and representations that help uncover the meaning of the operations. Finally, for each phase in building fluency for that pair of operations, we will provide general recommendations and suggested learning activities and games. As a reminder, the three phases as outlined in Section 2 are:

1. building foundational concepts of number and operations and learning to figure out simple facts with counting and visually based strategies;

2. learning *reasoning strategies* to determine facts more efficiently; and

3. achieving full fact fluency.

The Teacher's Role

When we enter classrooms where students are successfully learning their number facts, we see teachers listening and responding to student thinking in ways that encourage and promote their flexibility with numbers. This includes creating a classroom culture that encourages students to take risks as they attempt to put numbers together and pull them apart, acknowledges that mistakes are opportunities to learn, and celebrates the mathematical insights that come from sharing and discussing strategies for operating with numbers. When this classroom culture is in place, there is a buzz in the room, with children doing, thinking, talking, and listening like mathematicians making sense of their work. As you read through this section, consider that when a teacher listens to children's emerging ideas and provides questions that probe and stretch their thinking further, they are also building a math classroom culture of young mathematicians at work.

In the past, some educators have considered teaching practices for building conceptual understanding of number and operations to be distinct from teaching practices for supporting fact fluency. In this view, for example, a teacher would ask probing questions when students are working on a problem that requires critical thinking and then discuss the problem as a class. In contrast, when students are working on basic number combinations, that teacher would tell the students to work independently and silently. This is a false dichotomy: both complex problems and early practice with number combinations require mathematical sense-making and strategic choices for solution paths. As noted in Section 2, this is important for all students, including those with math difficulties.

The role of the teacher in building students' number sense and fluency with number combinations includes (1) setting learning goals and posing "just-right" problems, (2) asking questions to help students build on what they know as they work together to solve problems, and (3) using different students' ideas and strategies in a whole-class discussion that builds connections across these ideas and strategies. These approaches are consistent with the design of a student-centered lesson (Smith and Stein 2011) that supports student thinking and reasoning and helps build the strong conceptual foundations that are important for supporting fluency.

Let's look more closely at each aspect of the teacher's role and consider how to plan for this kind of instruction.

Setting Learning Goals and Posing "Just-Right" Problems

Understanding specifically what students know and can do allows a teacher to use a learning progression to identify a goal and pose a "just-right" problem. For instance, a foundational idea is cardinality, the idea that the last number counted represents the total quantity of the set, so an activity that involves counting and stating "how many" would be appropriate. When students are beginning to use and apply the reasoning strategy of doubles plus/minus one ("I know that 3 + 3 = 6, so 3 + 4 equals 1 more than 6 . . . 7"), we should choose problems that encourage them to use that strategy. In later grades, students first think about multiplication as repeated addition, but can then learn to apply reasoning strategies for that operation ("I know that 5 × 2 = 10, so 5 × 4 is double, 10 + 10, it's 20!").

When thinking about a goal for developing fluency, consider a learning goal rather than just a performance goal. A performance goal focused on fluency might be the following: "Students will be able to complete twenty basic facts accurately and quickly." On the other hand, a learning goal might state, "Students will be able to solve addition facts by applying a reasoning strategy such as doubles plus one." While you may want students to demonstrate their learning during the lesson and perhaps on an exit ticket at the end of the lesson, student learning rather than student performance should be the driver. This can support a culture of sharing and risk-taking rather than one of anxiety or unproductive competition.

Asking Questions

What do you notice?
Can you tell me the story in your own words?
What strategy can you use to solve this problem?
Does this problem remind you of another problem?
Can you draw a picture to help solve the problem?
What are you trying to find out in this problem?
Is your answer reasonable?

These simple questions can help empower students to look for and explore arithmetic properties, patterns, and relationships. They communicate that students have ideas worth talking about. While planning for a set of problems, teachers can anticipate what key ideas and strategies might arise and plan questions that might help students if they are stuck. For example, if a student says, "I don't know what to do," and the teacher asks, "can you tell me something you know about the problem?" When students are working on a problem, asking about their thinking and their solution strategies can help scaffold their learning and help them consolidate it. Such questions also encourage students to use what they know and notice in order to solve problems and develop fluency with different number combinations.

It can be challenging to restrain ourselves from correcting students when they make mistakes, but here too, questions can help. We have found that students often self-correct if given the opportunity to explain their thinking. Sometimes asking students "How do you know?" is sufficient to uncover both what they know and what is incorrect.

Using Students' Ideas and Strategies to Build Connections

After students have worked independently and in small groups, gather the whole class to facilitate a discussion based on your learning goal and what strategies, ideas, and struggles you have observed. For example, after students play a game that involves adding two numbers, a teacher might ask questions that extend learning (Teacher: "Can you always 'turn around' the addends and get the same total? What happens in subtraction?").

Here are some other questions that encourage students to discuss their mathematical thinking:

How did you solve the problem?
Can you explain your classmate's strategy?
How do you know that works?
Do you agree or disagree with _____ ?
Does that always work? Is that always true?

Discussions at the end of a lesson can help students consolidate their learning and push their mathematical understanding and use of efficient strategies. Remember the research presented in Section 2 that noted that the number of different strategies students use can predict their later learning. By encouraging student-to-student talk, you are also encouraging them to use their own "funds of knowledge" to do math and communicate their thinking.

Using questions that help students articulate their thinking so it can be more readily unpacked, rather than simply telling them the reasoning, helps students deepen their understanding, be more likely to persevere, and be more ready for their next challenge.

General Suggestions for Building Fact Fluency

How can we organize math class so that students have opportunities to build foundational concepts; explore, apply, and discuss reasoning strategies; and engage in robust, distributed practice to achieve fluency? In this section, we'll look at some key aspects of instruction that apply to all operations.

In general, we suggest a three-part lesson structure: first, a brief launch where the teacher poses the problem(s); second, an explore time where students are working independently and in small groups solving problems while the teacher circulates to observe, listen, and ask questions about student ideas and strategies, including misconceptions; and third, a summary or wrap-up where the teacher chooses selected strategies and ideas that emerge from the students to analyze and discuss. We will also look, later in this section, at other ways to provide opportunities for students to build fluency.

The Importance of Planning

There are three stages of effective planning. First, identify the learning goal of the lesson and how you might convey it in a student-friendly way. Second, do the math yourself. Importantly, anticipate how students might approach the problem, including ways they might get stuck, language or directions that might be confusing, and strategies and representations they might use. For each of these, consider what feedback you might give to move the learning forward, including questions you might ask to elicit further student thinking. This will help when you circulate while students are working independently and in small groups. Third, think about what you might focus on during a wrap-up discussion, while knowing this may change in response to what you actually observe and hear when students are doing the work.

Models and Representations

As you plan for activities and prepare for discussions that support the development of fluency, consider how to use models and representations. While visual representations are the most prominent, there are other kinds of representations, as shown in Figure 3–1.

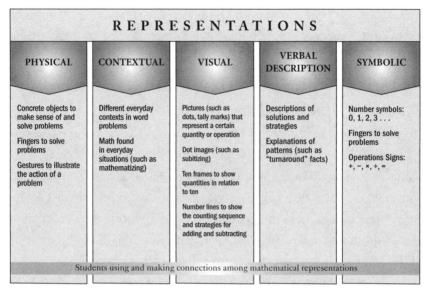

REPRESENTATIONS

PHYSICAL	CONTEXTUAL	VISUAL	VERBAL DESCRIPTION	SYMBOLIC
Concrete objects to make sense of and solve problems Fingers to solve problems Gestures to illustrate the action of a problem	Different everyday contexts in word problems Math found in everyday situations (such as mathematizing)	Pictures (such as dots, tally marks) that represent a certain quantity or operation Dot images (such as subitizing) Ten frames to show quantities in relation to ten Number lines to show the counting sequence and strategies for adding and subtracting	Descriptions of solutions and strategies Explanations of patterns (such as "turnaround" facts)	Number symbols: 0, 1, 2, 3 . . . Fingers to solve problems Operations Signs: +, −, ×, ÷, =

Students using and making connections among mathematical representations

Figure 3–1. Different types of representations. From *Principles to Actions* (National Council of Teachers of Mathematics 2014, 25) (adapted from Lesh, Post, and Behr 1987).

For example, as noted in Section 2, young children often begin to understand number and simple operations through story contexts. As students connect multiple representations, they have opportunities to strengthen their mathematical understanding, apply higher-level strategies, and become more fluent with basic number combinations. For instance, a teacher posed a typical problem: "Maya had 4 pennies. Then her sister gave her 2 more pennies. How many pennies does she have now?" A student solved it using a drawing with a row of 4 circles and then 2 circles below. When the teacher asked her how she knew that

See Section 2, page 28, for more on the importance of using fingers as representations.

4 + 2 = 6, the student held up her hands (4 fingers up on one hand and 2 on the other) and said, "See, 4, 5, 6." She responded to a word problem (contextual), solved the problem with a drawing (visual), and confirmed it by a statement (verbal) and a physical action (acting out the solution with her fingers).

We suggest specific representations for each pair of operations in the two sections about addition/subtraction and multiplication/division that follow.

Learning Opportunities and Activities

To give students a full set of experiences to develop fluency, use different types of learning activities during math class as well as at other times in the day. These include word problems, games, instructional routines, digital tools, and everyday situations. While math can be integrated into centers or linked to thematic learning, discrete time for math should begin in preschool and grow longer in duration as children move into kindergarten and the elementary grades.

In the rest of this section, we will take a closer look at other ways to build math fluency: we will see how to use games effectively, consider the use of instructional routines and digital tools, and, finally, learn how to partner with families to support their children's learning.

A Closer Look at Using Number Games to Build Math Fact Fluency

In Section 1, we discussed how playing a game might involve just following a set of directions and not necessarily affect math fact fluency. How can we shift our use of games so that they become contexts for learning and practice that build foundational concepts, provide opportunities for reasoning, and allow for practice to achieve fluency? Here are some recommendations:

Selecting and Planning for a Game
- Select a game that matches a learning goal you have for your students. You may want to consider how it can be adapted for a range of learners.
- As part of planning, play the game. Note if the directions to the game are complicated and consider how you will share the directions to the game.

- Plan how you will group students. Partner games are designed to encourage students to use their peers as resources to check each other's thinking and accuracy. If it is a game where one student wins, you may need to downplay the competitive aspects of playing and highlight what students can learn and how they can work as partners or small groups.
- Anticipate what you might ask or say to support the use of efficient strategies or help students when they get stuck.

Introducing a Game
- Convey the learning goal(s) of the game in a student-friendly way.
- Ask a student to be your partner to model the game with the whole class. As you share the directions, ask your student partner and others about the next moves they might make.

As Students Are Playing the Game
- Circulate as the game is being played, and observe, ask questions about the strategies students are using, and check for computational accuracy. In addition, you can consider what ideas about the game you want to share and discuss during the whole-group summary discussion.
- When appropriate, students can record mathematical equations or representations while playing games to help consolidate their learning and provide a window into their understanding and fluency.

Wrap-Up: Emphasizing the Learning Goal of the Game
- After playing the game, facilitate a summary discussion where students can reflect on what they learned and practiced, strategies they used, challenges they had, and how they might play the game the next time.
- Play games more than once to help students practice their math skills and apply strategies as they move toward fluency with their combinations. Games also provide a wonderful

home-school connection that encourages families to support their children in an engaging way.

An Example: Unpacking Toss the Chips

In the *Investigations 3* curriculum (Pearson 2017), kindergarten students play Toss the Chips, a game that uses two-color counters, red on one side and yellow on the other. Children toss a given number of counters onto a mat and record how many come up red and how many yellow. The game is designed to allow children to decompose a given number in many ways, record each combination, and begin to notice patterns and properties related to those combinations.

What mathematical ideas might arise when playing Toss the Chips over time? What questions might you ask to highlight those ideas? Some examples to explore are:

- Multiple decompositions can exist for a single number.
- Commutative property ($2 + 3 = 3 + 2$, $a + b = b + a$)
- Using doubles and doubles plus/minus one (($N + N$) +/– 1)
- An equation is not always about the answer ($2 + 4 = 6$, $6 = 4 + 2$).
- What happens when you add zero to a number?
- Recognizing the patterns and relationships of the totals ($5 = 0 + 5$, $5 = 4 + 1$, $5 = 3 + 2$, $5 = 2 + 3$)

How do you know whether your mathematical learning goal was met after your students play the game? Consider how students learn and understand different reasoning strategies, such as applying the commutative property ($a + b = b + a$; for example, $5 + 3 = 3 + 5$) to solve new problems. As you are circulating around the room, you might notice that one child has recorded both 5 red chips and 3 yellow chips and 3 red chips and 5 yellow chips for the target number of 8. At the end of the activity, you can gather the students to ask them how these two combinations are different and how they are the same. You can extend the conversation to whether this works for all numbers.

While many students need time to feel secure with the idea that order doesn't matter in addition, the context of Toss the Chips can make for a compelling discussion. It is important to remember that students do not need to know the term "commutative property" but do need to understand and be able to apply it.

Thoughtful Instructional Routines

Instructional routines are short (5–15 minutes) classroom activities that have a consistent structure with clear mathematical goals. Because instructional routines are brief, they work well as distributed practice throughout the year. They can be used as a warm-up to the lesson or as a transitional activity during the day. We have included examples of specific instructional routines that help support flexibility, reasoning skills, and fluency in the lists of activities for each of the operations.

> Sometimes students use hand gestures to show their understanding of what is happening or use the words "turnaround facts" when applying the property to a new number.

Digital Tools

There are many, many resources available online, both free and through purchase and subscription, that are targeted to support fluency in all four operations. Given the range of offerings, how do you choose and implement programs that can support fluency? Here are some considerations:

- Avoid programs that are old-fashioned worksheets dressed up online. If they merely present a series of computational fill-in-the-blank problems without engaging students in any cognitive work, keep looking.
- Consider the ways a program supports students in using arithmetic properties, patterns, and relationships. Two free online programs are Open Middle problems (www.openmiddle .com) and Splat! (www.stevewyborney.com). Open Middle problems highlight that there are many ways to solve a problem and emphasize strategic thinking. In Splat! students see a given number of dots, then a splat covers some of the dots and students need to figure out how many dots are under the splat. Guiding questions are provided for the slides. There are other

routines available on the site, including an estimation clipboard and ideas for working on subitizing (see Section 2, page 33).

- As noted in Section 2, research has shown that some speed-related practice can help achieve fluency. Digital tools that give speed practice should be used only when students have foundational concepts and reasoning strategies in place for the targeted operations and range of numbers (e.g., addition and subtraction combinations within 10).

Everyday Opportunities for Math

Everyday situations provide real-life opportunities for seeing math in action (known as mathematizing) and for solving problems. For example, as students look around the room, they may see window-panes set in three rows with four panes in each row. This provides an entry into a discussion about multiplication situations and encourages them to keep noticing the math in their own world.

Family Engagement and Collaboration

Families are their children's first teachers and can be powerful partners in supporting math fact fluency. Like many of us, family members may have experienced "drill and kill" instruction and rote memorization, and they may think that they should speed test their children on isolated facts. Instead, you can help them be active partners in their child's development of number sense and fluency that are based on understanding. Consider ways to partner with and engage families in supporting students:

- Share simple games that provide distributed practice for students. Provide game directions and resources that family members can bring home to play with their children.
- Host a math night that highlights how fluency develops, and have families try out different activities and games. Encourage families to discuss what they notice and reflect on what they are learning.
- Provide homework guidelines that help frame a positive role for families through questioning and an inquiry approach. For example, provide sample questions that families can ask when

children are having difficulty. A general guide from PBS that families can use to help with homework can be found at www.pbs.org/parents/education/going-to-school/supporting-your-learner/homework-help/.

- Host a math celebration breakfast and have children share their learning in an interactive way. Math journals and student work can be on display.
- Provide resources that help families understand the progression of ideas and skills. For example, the Council of the Great City Schools has Parent Roadmaps in Mathematics that show the key math ideas for each grade, as well as what comes before and after that grade. They are available at the Resources and Tools tab at www.cgcs.org/Page/244. The National PTA has a Parents' Guide to Student Success for each grade, which provides a concise resource for understanding the standards and how to support one's child. They are available at www.pta.org/home/family-resources/Parents-Guides-to-Student-Success/.
- Another useful link for parents is www.pbs.org/parents/education/math/. It has everyday activities for families, math milestones by age band, and advice and tips for making math fun.

Developing Fluency with Addition and Subtraction

In this section, we first explore the models and representations that support fluency with addition and subtraction. We then consider general recommendations and specific learning activities for each of the three phases:

1. building foundational concepts of number and operations and learning to figure out simple facts with counting and visually based strategies;
2. learning *reasoning strategies* to determine facts more efficiently; and
3. achieving full fact fluency.

Models and Representations for Addition and Subtraction

The following models and representations help students make sense of and solve a problem, notice relationships between numbers, and communicate their ideas as well as understand the ideas of others.

Dot Images

The importance of perceptual subitizing was discussed in Section 2. As a review, look at the dots in Figure 3–2 and think about how many you see and how you know.

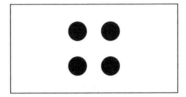

Figure 3–2 Dot pattern

Did you need to count the dots or did you instantly see four? Subitizing means to "instantly see." The next step is to build students' ability to conceptually subitize by recognizing two groups or more and quickly combining them. Look at Figure 3–3 and again think about how many dots you see. How might you explain how you saw them? Can you think of other ways that someone might figure out how many dots there are?

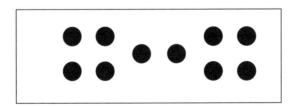

Figure 3–3 How do you see these dots?

We will discuss how to provide subitizing activities in the section on recommended activities and games.

Five and Ten Frames

Five and ten frames provide an organizing visual structure that high-lights ten as a landmark or benchmark number. You can initially use a five frame for small numbers, then a ten frame for numbers 0 to 10, and two ten frames for numbers 11 to 20. To convey the importance of 10, many teachers say something like "My favorite number is 10—it is so friendly and helps me solve lots of problems!"

Students can use five and ten frames to see the relationship of a given number to either 5 or 10. By using these benchmark numbers, children can see a small group of dots by subitizing them. Both of the frames in Figure 3–4 show the quantity 3, one in the context of 5 and one in the context of 10.

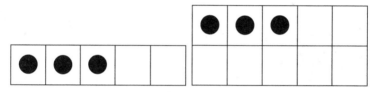

Figure 3–4 **Five and ten frames for the number 3**

Sometimes students spontaneously see the 3 in relationship to the 5 spaces within the frame. "I see 3 dots and 2 missing dots." If they don't, and as appropriate, ask how many you need to get to 5 (or 10). As students have more experience with this type of question, they begin to understand the relationship between addition and subtraction and how subtraction can be viewed as a missing addend (3 + ? = 5). This is a critical idea in developing fluency. Once students understand that subtraction is closely related to addition, they can use their knowledge of addition combinations to quickly solve subtraction problems.

As students work with different arrangements, they begin to see equivalence between certain combinations. For example, students may see the image in Figure 3–5 as 3 + 4, 6 +1, or 2 + 2 + 2 +1.

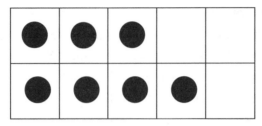

Figure 3–5 Ten frame for 7

Two ten frames can also support students in developing the strategy of break-apart-to-make-ten (BAMT). Students often notice that they can decompose one addend to make a 10 with the other addend and quickly find the total. For example, given a ten frame with 8 dots and another with 6 dots, students will break apart the 6 into 4 and 2, adding the 2 dots to the 8 and declaring, "10 + 4 = 14!"

> As mentioned in Section 2 (see page 29), fingers are reliable, but they are not the most efficient tool. Support students in solving problems without fingers as they become ready.

Don't Forget Fingers!

Fingers are the reason for our base ten system and they are always handy. They are also the first way for kids to think abstractly—fingers represent something else, such as numbers or objects. Fingers can help a student apply and show the strategy of counting on. For example, when solving a problem such as 7 + 5, a student might hold the first and largest addend, 7, in his head and then count on using 5 fingers, "7 ... 8, 9, 10, 11, 12". Teachers can connect fingers to other representations, such as dot images. As mentioned in Section 2 (see page 29), fingers are reliable, but they are not the most efficient tool. Support students in solving problems without fingers as they become ready.

Counters

The research in Section 2 highlighted the importance of the thoughtful use of manipulatives, including counters. Very young children benefit from the use of concrete objects that help them make meaning of the task. They can represent the action of a word problem and keep track of counting moves.

An example of counters that can serve a specific purpose are two-color counters (often red on one side and yellow on the other), which can show the different combinations that a target number can have. Additionally, materials such as popsicle sticks or stacking cubes help students make units of tens and can support fluency with facts between 10 and 20 as well as the development of early place value understanding. It is important to allow students to move from using concrete objects to abstract reasoning as they establish successful strategies and apply them to new situations.

Number Lines

A number line is a useful representation of our number system. It helps students visualize the counting sequence and illustrate their strategies for adding, subtracting, multiplying, and, to some extent, dividing. Each point on the number line represents a real number. While a number line can start or end with any number, young children, usually beginning in first grade, see and use the number line starting at zero and growing in increments of ones (see Figure 3–6).

Figure 3–6 Number line

As students move from exploring the number line as a counting tool to thinking about reasoning strategies, they can use the number line for recording different strategies. For example, when students solve the problem 5 + 3, they might show the counting on strategy using the number line. Knowing that they do not need to begin at 0, they may begin at 5 and then make 3 jumps of 1, as shown in Figure 3–7.

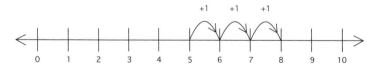

Figure 3–7 Three jumps of 1 on the number line

If students are solving for 9 + 6, they may use the break-apart-to-make-ten strategy and show their thinking on the number line. They can break apart the 6 into 1 and 5, add the 1 to the 9 to make 10, and then do a jump of 5 to get to 15. Figure 3–8 is an example of an open number line where students make the numbers and tick marks to match their thinking.

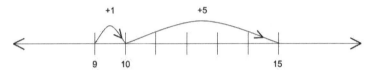

Figure 3–8 Open number line for 9 + 6

With a subtraction problem such as 9 – 5, students can use a subtracting back strategy or think about the problem as a missing addend and add to solve. The first representation in Figure 3–9 represents the take away model of subtraction. The second highlights the relationship between addition and subtraction and highlights the distance between two numbers.

9 - 5

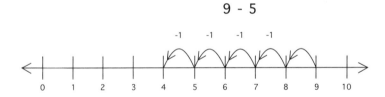

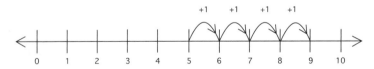

Figure 3–9 Two strategies for 9 – 5

Again, encourage students to connect their number line representation to an equation.

Tape Diagrams and Number Bonds

Tape diagrams and number bonds are representations that show part-part-whole relationships and provide opportunities for students to compose and decompose numbers in many ways. They can be helpful for representing word problems such as "how many of each?" ("I have 5 apples. Some are red and some are green. How many of each could I have?") and comparison problems ("Rosa is 8 years old. David is 6 years old. How much older is Rosa than David?"). These types of word problems do not involve action and instead present static images.

In the tape diagram, the addends are tapes and the total (indicated by a brace) is the combination of those tapes (see Figure 3–10). It can also be used to represent different unknowns in word problems.

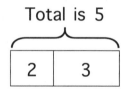

Figure 3–10 Tape diagram

The number bond diagram represents the total at the top and the addends at the bottom. It can illustrate the different unknowns (such as total unknown, start unknown, or change unknown) as it highlights part + part = whole or whole – part = part (see Figure 3–11).

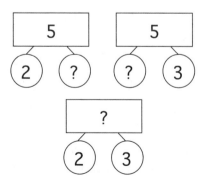

Figure 3–11 Number bonds

Sometimes the number bond is shown in an abbreviated form, as in Figure 3–12.

Figure 3–12 Abbreviated number bond

Rekenreks

The rekenrek (see Figure 3–13), designed by Adrian Treffers, is a kinesthetic and visual tool consisting of two rows of ten beads with each row made of five red beads and five white beads (or other contrasting colors). It uses the numbers 5, 10, and 20 as anchors for counting, adding, and subtracting. There are individual rekenreks that can be made or purchased for students and larger class rekenreks for teacher use. Ways to use rekenreks are included in the sections on recommended activities and games that follow. For more information about this tool, visit www.k-5mathteachingresources.com/Rekenrek.html.

Figure 3–13 Rekenrek

Phase 1: Building Foundations and Initial Strategies for Addition and Subtraction

Very young children have mathematical experiences rooted in everyday life. The word *more* is often one of the earliest words that babies learn. A toddler can want two cookies because she has two hands or notices that when she eats one cookie she has less.

Children benefit from varied and robust opportunities to count, subitize, compare and order, join quantities together to make a new set, or take apart a set. As students begin to understand that quantities and numbers can be composed (put together, joined) and decomposed (taken apart, taken from), they develop a foundation for mathematical success in the elementary grades and beyond. Later, this ability to compose and decompose supports students' fluency when learning about equal groups, supporting them to multiply and divide flexibly, efficiently, and accurately.

As you read the following classroom vignettes and suggestions, consider how early experiences with counting can provide a strong foundation for adding and subtracting. When preparing to use the following activities and games, consider the following questions:

> *What mathematical learning goal or goals are embedded in the specific activity?*
> *How might I communicate the goal(s) to my students?*
> *What purposeful questions can I ask that can extend student thinking related to the goal(s)?*

Vignette: Snapshots and Subitizing

Early in the school year in a prekindergarten classroom, a teacher has prepared three counters on a paper plate covered by a cloth. The students are sitting in a circle after singing a short counting song. The teacher asks them to remember when their families took a photo with their phone or a camera. She explains that they will be taking a photo or snapshot with their eyes of what they see. She makes a clicking sound like a camera and says, "Watch carefully—your eyes have to take a very quick picture!"

She tells them to use their fingers to show how many they see. "Ready?" The students nod. She uncovers the plate for two seconds and covers it again. Most children hold up 3 fingers, though a few appear hesitant and look at their friends. She asks several students how many they saw. "Let's look again." This time she leaves the plate uncovered and asks what students see.

The teacher repeats this activity using different quantities of counters up to 5. Each time she explicitly tells them to take a picture using their eyes and their mind (or brain). She is clear about the goal of the activity—to have students "subitize" or instantly see a small quantity, begin to understand the relationship between quantities and number words using different arrangements, and create an equivalent set with their fingers.

Later that year, a small group of students are working with the same teacher. She is holding a covered plate with 4 circle counters underneath the cover. She quickly lifts the cover and they quickly look at the image and say "4!" showing 4 fingers. She asks them how they know ("I used my brain." "I just saw them." "I counted super quick.") She then re-covers the counters with a small cloth, shows that she has another counter, and states, "I am adding 1 more counter. How many counters will there be now?"

This simple activity provides an opportunity for students to link counting on to adding. The teacher can repeat this activity during the school year using other small quantities of counters and adding or taking away 1 or 2 each time. Another time she plans to introduce subtracting to 0 ("I have 2 counters. I am going to take away 2—how many will I have?"). By practicing this activity with students over time with a variety of quantities, she is supporting students' later fluency with combinations within 5, which provides the foundation for fluency with higher numbers. She also wants to encourage them to begin to express what they see and know in different ways.

Begin by Counting Everything

"How many?" is an intrinsically interesting question for young children. There are lots of opportunities for counting in the classroom: shapes, people, pencils and crayons, snacks, and stairs, to name a few. Early counting experiences are the foundation for adding and subtracting. By kindergarten, students should have developed cardinality: an understanding that the last number named tells the number of objects counted. This is a milestone—without it, when adding two quantities, students will continue to count all instead of counting on. Students can also begin to see the counting sequence as their first algebraic pattern, $N + 1$ (without naming it with symbols, of course).

Ask students not only to count a given set of objects but to *produce* a given set: "Can you get me 3 pencils? Can you show me 6 counters?" And don't forget to count backward as well as count forward! This is a more difficult skill that also requires distributed practice and prepares students for subtraction. Gesturing with your hands in equal increments as you count up or down can support their understanding of the pattern of the counting sequence (adding ($N + 1$) or taking away ($N - 1$)), where each number counted is one more or one less than the previous number.

Math Is Everywhere

Use everyday situations to engage students in working with lots of counting and with simple addition and subtraction combinations. Snacks are a frequent setting for these conversations; children are often interested in how many crackers or apple slices they have. By using everyday situations, we help students begin to notice the relationship between an action (initially either adding to or taking away) and an operation.

Teachers can also take students for a neighborhood walk looking for images that can be used to tell a math story. This also supports student connections to their own community and is a wonderful experience to share with families.

Use Your Senses

See It: Quickly show different arrangements of small numbers using dot images and ten frames. Wait two seconds and then cover or hide the image. This teaches children to subitize and make a mental image of a small set rather than counting by ones (see page 23).

Hear It: While visual subitizing is most common and effective, we can also hear small sets of numbers. You can clap a small number of times and ask students how many claps they heard.

Move It: Show students a given number (written large on paper, an index card, or a whiteboard). Ask them what the number is, and then have them jump, clap, or stamp that number.

Show Me (with Rekenreks): Say a number or hold up a numeral card from 0 to 10, and ask the students to show the given number by moving the beads on their rekenreks with one push. For numbers 11 and 12, students use two pushes of the beads.

Number Talks

Number Talks is an engaging instructional routine initially developed by Kathy Richardson and Ruth Parker in the 1990s. It has continued to develop as a strategy to encourage students to make sense of mathematics, communicate mathematically, and justify their solutions. Teachers often use the following protocol:

1. The teacher provides the problem.
2. The teacher provides students sufficient wait time to solve the problem mentally. Mental solutions encourage students to use efficient strategies.
3. Students show a visual cue when they are ready with a solution. One visual cue is a thumb up against one's chest. This conveys readiness without the distraction of raising and waving one's hand.
4. The teacher calls for answers. They collect and record all answers—correct and incorrect.
5. The teacher calls on students to share strategies and justifications with their peers.

Initially you can use dot and ten frame images in supporting young students to recognize small quantities and then to be able to add one or two more to what they know.

Games

Snap It

Materials: a tower of ten cubes using one color

You can introduce this game to the whole class or in a small group.

Show the tower of cubes. Have students count aloud so they know that there are ten cubes. Dramatically say, "Snap!" and break the tower into two parts, hiding one part behind your back and showing and counting the remaining part. Then ask the students how many cubes are hidden and how they figured it out.

This is an activity that students can work on with partners and can be used to highlight the relationship between addition and subtraction.

One More, One Less

Materials: Number cards 1–5 or 1–10 and +1/–1 dice.
A recording sheet is optional.

Early games with dice or cards can support emerging addition and subtraction. Consider making or purchasing dice that have "+1" and "–1" instead of the usual 1 to 6 dots. In One More, One Less, a game in the Investigations 3 curriculum (Pearson 2017), students turn over a number card from a deck, roll a +1/–1 die to see if they should add or subtract 1, and then record their answer.

X-Ray Vision

Materials: Number cards 1–5, 1–10, or 1–20. You may want to include a 0 card.

This game is included in the *Building Blocks* program used in many pre-kindergarten classrooms (Clements and Sarama 2007–13). It encourages students to develop a "mental number line" and to see the relationships between numbers, and supports the strategies of counting on and counting back.

Place cards in order, face down. *If you use a 0, put that card face up.*

You can say, "I am thinking of the number 3. Can anyone find the number 3?" A student turns over that number card, often counting "1, 2, 3" to find the right one. Ask the student, "How did you know?" or "Can you explain your thinking?"

If a child guesses incorrectly, also ask them to explain their thinking. Often students can figure out their own mistakes. If the child can't, you can review counting from one if necessary or have the student ask a peer for help. This sends the message that mistakes are part of learning and also provides assessment information for you.

You may also say, "I am thinking of a number that comes next after 3. Can anyone find the number that comes next after 3?"

Continue to prompt students to find specific numbers, or numbers that come before, after, or between other numbers.

Students can work independently or in pairs, taking turns pointing to cards and using their "X-ray vision" to tell which card it is. Children turn their card over to show whether they are correct.

Counting Jars and Equivalent Sets

A counting jar holds a specified number of items for children to count. You can use the same jar all year. Begin by asking children to take out the items to count them, and then ask them to make a set of their own that has the same number of objects and record the amount.

Variations to the counting jar include the following:

Have students guess how many items before taking them out of the jar. Asking students to guess "how many" helps them develop a sense of magnitude and communicates the idea of a reasonable answer.

Have two or three jars to account for the range of quantities different students are working on to provide "just-right" challenges for everyone.

Comparing and Ordering

As children master the rote counting sequence, they can begin to order and compare numbers. Give students a mixed-up deck of number cards and ask them to put the cards in order from least to greatest. Depending on their counting fluency, the cards can be up to 5, 10, or 20 and can also include 0. They can then use counters to match quantities to each number. You can also give them a range of numbers not starting at 1 or 0 (for example, 5 to 10 or 7 to 15). You can ask questions such as "Which is more?" "Which is less?" and "Which comes first?"

Different Ways to Make a Number with Rekenreks

This rekenrek activity can be used to show different compositions of a given number. (For more information, see www.k-5mathteachin-gresources.com/Rekenrek.html.) Begin with only the top row of beads. You can cover the bottom row with a strip of paper or fabric. Slide the red beads to the left and the white beads to the right on the top row of the rekenrek. Choose a number to build. You can say, "Let's see how many ways we can build 5 by sliding beads from each side to the middle. What if I slide 3 red beads from the left and 2 white beads from the right? Does that make 5 beads? Can you think of another way to make 5?"

You can repeat this activity using the numbers from 1 to 10. Once children become familiar with using the top row, the class can begin to find combinations by using both the top and bottom rows. Students can also record the different ways they find to build the given number using drawings, numbers, and/or equations.

Literature Connections

Consider how to use counting books to engage students in thinking about growing patterns. Here is an example from a classroom.

A teacher is reading the story The Napping House *by Audrey Wood to young children in the meeting area. It is a rhyming book with sweet and amusing characters, human and animal, that one by one go to sleep on top of one another, with a grandma at the bottom and a flea at the top. With the turning of each page, the pile of sleepers becomes more precarious until finally each character awakes and falls off. The children have heard the story before and anticipate the rhyming words and the silliness. The teacher pauses periodically and asks how many characters are sleeping now and what would happen if one more joined them.*

There are many other similar stories that provide rich opportunities to think about the counting sequence.

General Recommendations to Support Building Foundations and Initial Strategies

See page 47 for suggested questions

- Provide lots of opportunities for students to represent, notice, and wonder about quantities and counting and relate counting to early addition and subtraction work.
- Use learning progressions and state standards to guide tasks. We recommend the *K–5 Progressions on Counting and Cardinality* and *Operations and Algebraic Thinking*, available at http://ime.math.arizona.edu/progressions/.

- Ask questions that assess reasoning and fluency while also stretching students' learning and thinking.
- Begin with small numbers to add or take from. Always build on what students already know and stretch their thinking just enough, such as adding or taking away 1 or 2. For example, if a student counted 4 teddy bears, ask how many more she would have if you gave her 2 more bears.
- As children become fluent with combinations of small numbers, begin to introduce combinations with larger numbers, perhaps posing a problem with one large number and one small number (e.g., 5 + 2) so that children begin to count on (5 . . . 6, 7).
- Have students use different representations: showing their fingers, drawing, acting out a story, making gestures that show action, and using movement ("Let's jump to 4").
- Find opportunities for students to notice, represent, and talk about the relationship between addition and subtraction.

Phase 2: Learning Reasoning Strategies for Addition and Subtraction

How might you subtract 98 from 100? Would you line up the numbers in vertical notation? Would you then try to take 8 from 0 and say it is impossible? (Mathematically, it is not—it just requires use of negative numbers.) Would you then "borrow" "one from the tens column" and later another "one from the hundreds column"?

Or might you use a reasoning strategy, perhaps counting on or counting back? "Oh, I know 98 is 2 away from 100 (98 is my starting number, 99, 100)," or "I can use addition—since I know my combinations of 10 (8 + 2 = 10), then 98 + 2 = 100, so 100 – 2 = 98." You can probably solve the problem mentally much faster than you can say its steps or grab a pencil to record the algorithm. You can solve the problem quickly because you are using an efficient strategy suited to a certain set of numbers.

As students learn reasoning strategies for solving single-digit addition and subtraction combinations, they are developing fluency: they are using accurate knowledge and concepts and strategies that promote their adaptive expertise, as defined in Section 2. As a reminder, effective strategies include counting on, doubles plus one, conceptual subitizing, and break-apart-to-make-ten.

Vignette: Number Talk

In a second-grade classroom, a teacher begins a Number Talk by writing the problem 17 − 9 on the board and asking students to solve it mentally. No hands are raised; instead, students show that they have solved the problem by a quiet thumbs-up on their chest.

The teacher asks for answers and notes them on a corner of the board: 8, 9. She calls on the student who answered 9, and asks him to explain his thinking. The student begins, "I counted back: 17, 16, 15 . . . ," and as he speaks the teacher records his countdown on the board. Suddenly the student realizes his mistake and self-corrects: "No, it's 8." The teacher asks him why he changed his answer, and the student responds that he counted the 17 and the 9 when counting back.

The teacher asks another student how she solved the problem, and she explains, "I added up. There's 1 more to 9 to get to 10, and then I know there was 7 more to get to 17. 1 + 7 = 8, so 17 − 9 = 8." Another child volunteers, "I did something like that, but I subtracted back from 17 to 10, which is 7, and then I knew 1 more back would get me to 9." One last child says, "I knew that 9 + 9 = 18, so 18 − 9 = 9. But 17 is 1 less than 18, so the answer is 1 less than 9; it's 8." The teacher asks someone else to explain that strategy in their own words. Many students hesitate, so the teacher asks them to turn and talk. She asks, "Do you agree that this works? Why or why not?" A discussion follows, and the teacher uses a number line to record the different explanations.

At the end of the talk, the teacher names and reinforces the strategies students used. She states that together they have solved a subtraction problem by counting on, adding up, subtracting back, and using what they know about doubles. One quiet girl raises her hand and says, "Talking about numbers is good for life. You learn to keep trying and working."

As you prepare to use the following activities and games, remember to think about learning goals and purposeful questions you can ask.

Number Talks and Reasoning Strategies

The protocol for Number Talks requires that students solve problems mentally and share their strategies and justifications. This encourages students to build on what they know rather than on memorized procedures and switches the focus of the conversation from the correct answer to *how* you got the answer. Often students come up with several ways to solve a problem and learn from one another. As students learn a strategy such as doubles plus one or break-apart-to-make-ten, and continue to use it efficiently, they are moving toward achieving fluency.

During this phase, you can use dot and ten frame images to encourage conceptual subitizing and addition and subtraction expressions. Both help students develop reasoning strategies.

Games

Tens Go Fish

This game is from *Investigations 3* (Pearson 2017).

Materials: Decks of number cards. Include multiple copies of each number from 1 to 9.

Each player is dealt five cards from the number card deck.

Each player looks for pairs from their cards that make 10. Players put down the pairs of cards that make 10, and they draw new cards to replace them from the deck.

Players take turns asking each other for a card that will make 10 with a card in their own hands.

If a player gets the card, they put the pair down and pick a new card from the deck.

If a player does not get the card, they must "go fish" and pick a new card from the deck.

If the new card from the deck makes 10 with a card in the player's hand, they put the pair of cards down and take another card.

If a player runs out of cards, they pick two new cards.

A player's turn is over when no more pairs can be made that make 10.

Roll and Record

This game is from *Investigations 3* (Pearson 2017).

Materials: Dice or number cubes and a recording sheet.

Students roll two dice or number cubes and add the numbers. They write the total on a recording sheet. The game is over when one number column is full (see Figure 3–14).

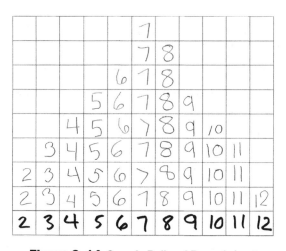

Figure 3–14 Sample Roll and Record sheet

By using number cubes without dots, students can develop strategies that do not involve counting all the dots. To support this shift, ask questions about how they solved the problems and share strategies during a whole-group discussion.

You can also play Roll and Record Subtraction by using a 7–12 number cube and a 1–6 dot/number cube. Have students play independently and roll the two cubes. They then subtract the smaller number from the larger number. They record the answer on their recording sheet. The game is over when one column is full.

Hitting a Target Number

Materials: 0–10 number cards; target cards 11 to 20.

Students must add or subtract two or more of the five number cards to arrive at the "target" number. Students may discard one or more cards if they cannot find a combination.

As students find the different number combinations for the "target" number, they explain their strategies to one another and check each other's solutions. They then write the equations on the piece of paper.

Dots Match

Materials: Cards with matching dot images.

Show four dot images on cards, all but one of which have the same number of dots (see Figure 3–15 for an example). Ask children which card does not belong. Ask them how they know. Students can also use these cards in pairs or small groups. This game emphasizes the different arrangements for a given number.

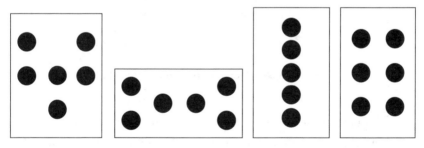

Figure 3–15 Sample cards for Dots Match game

Building Missing Addends

Materials: Rekenreks and number cards 1–10 or 1–20.

This activity involves rekenreks and helps students connect the idea of missing addends to subtraction. (See www.k-5mathteach ingresources.com/Rekenrek.html. for more information.)

Students work as partners and take turns choosing a card (the target number). The first partner moves some beads over on the top row of the rekenrek as the first addend, and the second partner needs to figure out how many more are needed to reach the target number and then move that many beads. For example, if a student picks the number 6 card, they may move 4 beads on the top row and their partner will need to move 2 beads on the bottom row. Partners make the number in as many different ways as possible.

General Recommendations to Support Learning Addition and Subtraction Reasoning Strategies

- Listen to children's thinking as they solve a problem. Have their ideas as well as their misconceptions guide your next set of questions.
- Encourage students to use known facts to find unknown combinations.
- Value mistakes as a way to assess what students know and are able to do and as a way to build their perseverance.
- Use talk moves such as wait time, turn and talk, and having students restate or add on to other students' explanations.
- Highlight that students can use reasoning strategies ("What strategy can you use to solve this problem?"). You are also sending the message that mathematics makes sense and problems can be solved in multiple ways.
- As students are exploring composing and decomposing quantities, connect their ideas to visual representations. You can also connect these representations to equations that match how they solved the problem.
- Provide names for the strategies that students can own, understand, and apply, such as counting on, break-apart-to-make-ten, doubles plus/minus one, or adding up (to solve a subtraction problem).

Phase 3: Achieving Fluency with Addition and Subtraction Facts

Remember the idea of adaptive expertise, the meaningful knowledge that can be flexibly applied to new, as well as familiar, tasks (Baroody and Dowker 2003)? Students achieve fluency when they have accurate knowledge, concepts, and strategies that they can apply flexibly and efficiently to both routine and new tasks.

Distributed practice is key to achieving fluency. While this means brief but regular practice, even at this level it does not mean "drill and kill" worksheets. Rather, students solve problems, play games, and engage in different activities. Memory and concentration games where students match a number to an expression (e.g., 12 matched with 15 – 3) can be useful and fun. At this stage, students are applying place value understanding and relationships between numbers to "know from memory" all sums of single-digit numbers. They are also solving a subtraction problem as a missing addend problem.

Ways to Make a Number

This instructional routine is from *Number Sense Routines: Building Numerical Literacy Every Day in Grades K–3* (Shumway 2011) and allows students to generate different ways to make a target number.

First, select a target number that aligns with the fluency combinations students are currently working on. You may choose a number up to 5, 10, or 20. Depending upon your goal, you may ask for a specific operation ("Let's think of two addends that make the number 20," or "Can someone think of a subtraction equation that makes 10?"). Gather many solutions so that students experience different ways to compose and decompose numbers.

Games

Addition and Subtraction Concentration

Make (or have students make) a set of cards with expressions (addition and/or subtraction) on half of the cards and their matching totals/differences on half of the cards. Place them face down in rows and columns. Students take turns, as in the traditional concentration game, finding matching pairs.

Fluency Board Game

Materials: Game board, dice, and color counter for each player.

Create a game board with start and finish spaces and with different addition and subtraction expressions in each of the other spaces. Students can also decorate and add fun touches (bridges, spaces with extra turns, etc.).

The first player rolls a die and moves the matching number of spaces. That player then needs to solve the expression and state the complete equation. The second player checks to make sure it is accurate. It is then the second player's turn to roll the die. Players continue to take turns until a player reaches the finish line.

True or False? (or Valid Inequalities?)

This game is from *Illustrative Mathematics* (see www .illustrativemathematics.org/content-standards/tasks/466).

In this activity, students decide if the equations are true or false and explain their thinking. It can be done as a whole group, by students working in pairs, or as a worksheet. This activity helps students understand that equality is about each side having the same value, while helping them achieve and demonstrate their fluency. Here are some examples of true and false equations:

$2 + 5 = 6$	$3 + 4 = 2 + 5$	$8 = 4 + 4$	$3 + 4 + 2 = 4 + 5$
$5 + 3 = 8 + 1$	$1 + 2 = 12$	$12 = 10 + 2$	$3 + 2 = 2 + 3$

General Recommendations to Support Achieving Fluency with Addition and Subtraction Facts

- Choose specific ranges of numbers based on fluency goals (0–5, 0–10, 0–20), and group-related facts.

- Choose number combinations that connect to or relate to specific related strategies (+/–1 or 2; +/–9 or 10).
- Create recording sheets or give students paper to record equations. When appropriate, ask them to notice patterns they see.
- Provide several opportunities for brief fluency practice during the week. Rather than providing long sessions on adding and subtracting facts, distribute practice with a number of short sessions over a longer period of time.
- Support connections across operations by eliciting student thinking about the relationship between addition and subtraction. Encourage students to solve by adding up. This reduces errors and also makes knowing basic subtraction facts much easier.

Developing Fluency with Multiplication and Division Facts

As students begin to work on multiplication and division within 100, they need to have many experiences composing and decomposing numbers using *equal* groups. As with addition and subtraction, they can build on what they know and begin to develop strategies toward achieving fluency.

According to the Common Core State standards, third-grade students are expected to both build conceptual understanding about multiplication and division as well as develop fluency with the single-digit numbers:

> Fluently multiply and divide within 100, using strategies such as the relationship between multiplication and division (e.g., knowing that $8 \times 5 = 40$, one knows $40 \div 5 = 8$) or properties of operations. By the end of Grade 3, know from memory all products of two one-digit numbers. (National Governors Association and Council of Chief State School Officers 2010, 23)

To "know from memory" is not the same as to rotely memorize. Students may have automaticity with some combinations (particularly with 2s, 5s, and 10s) but may need to quickly use derived facts to solve others ("I know $6 \times 6 = 36$, so 6×7 is one more group of 6, or 42").

Models and Representations for Multiplication and Division

For multiplication and division, begin with concrete and pictorial representations of equal groups and skip counting on a number line. The array and area models then follow as key representations to build understanding of multiplication and division.

Number Line

The number line can serve as a transitional representation between addition and multiplication. The related actions of repeated addition and skip counting are highlighted on the number line in Figure 3–16. A teacher may ask, "How many jumps of 3 do you see?" to help students connect this representation to multiplication.

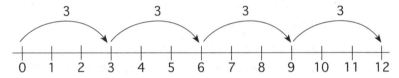

Figure 3–16 Number line for multiplication

Array and Area Models

An array is a group of objects arranged in rows and columns. While one convention with the array model is that the number of rows tells how many groups there are and the number of columns tell the size of each group, the array model can also highlight the commutative property of multiplication ($4 \times 6 = 6 \times 4$). The product is the total number of objects in the array (see Figure 3–17).

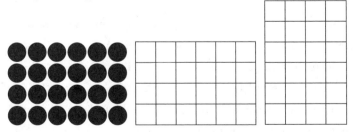

Figure 3–17 Sample array and area models for 4 × 6 and 6 × 4

As students internalize the square units of the array and area models, they can more efficiently think about the dimensions of each model and also relate multiplication to geometric measurement (see Figure 3–18).

6

Figure 3–18 Sample area model

Phase 1: Building Foundations and Initial Strategies for Multiplication and Division

Deep conceptual understanding and procedural fluency with addition, and to some extent subtraction, provide the foundation for multiplication and division. Students begin to understand the operations of multiplication and division through different contexts and representations. As we have seen, through foundational work with addition and subtraction, students notice and eventually justify how numbers behave; for example, they come to understand that order does not matter with addition but does with subtraction. That ability to notice, wonder, and move toward justification and generalization helps students develop fluency for multiplication and its inverse, division.

Of course, students have probably had experience with multiplication and division situations prior to the more formal work that can begin in third grade:

- That familiar phrase of "sharing is caring" can translate into preschoolers sharing snacks, such as making equal groups of crackers. Initially, division is conceptually easier for students because the "total" (dividend) is given, the divisor is the number of classmates/friends and oneself, and the quotient is easily countable and sometimes yummy.

- Equal groups are all around us as we buy juice at the grocery store; purchase supplies for the start of school; and arrange tables, chairs, and sets of crayon boxes in the classroom or at home. Children can begin to understand equal groups in everyday contexts and can also begin to see the array/area model in a drawer arrangement in their bedroom or a set of windowpanes in their classroom.

- In kindergarten, students learn to count by ones and by tens ("10, 20, 30, 40 . . ."), often expressing glee when they get to 100. Counting by tens helps young children understand the pattern of our base ten system (after 20 comes 21, 22, 23 . . . ; after 30 comes 31, 32, 33 . . .). In first grade, students can begin to see "ten ones" as "one ten." As students see that 2 groups of 10 equals 20, a foundation for later multiplication is being built.

- By the end of second grade, students are expected to be able to skip count by twos, fives, and tens. Skip counting is about being able to "chunk" or make equal units of numbers. With this ability, the road to fluency with multiplication and division is a much smoother one.

When preparing to use the following activities and games, consider how to assess students' early ideas about multiplication and division. As always, think about purposeful questions you can ask to extend student thinking related to the learning goal(s).

Quick Images

As you look at Figure 3–19, consider how many dots you see and how you know.

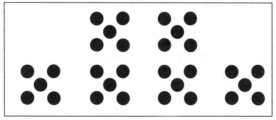

Figure 3–19 How many dots do you see?

This activity is sometimes referred to as "quick images," as a teacher flashes an image for about three seconds so that students capture a mental image that relies on conceptual subitizing. In this case, if students see a given number of groups, each of equal size, they can figure out the number of dots fairly efficiently. This activity can support students in moving from thinking additively to thinking multiplicatively. For example, one student might see the groups of 5 dots and add them together $(5 + 5 + 5 + 5 + 5 + 5)$, another student might skip count by groups of 5 (5, 10, 15, 20, 25, 30), and a third student might think, "I see 6 groups of 5, so $5 \times 6 = 30$." By sharing, recording, and connecting these different solutions, a teacher can help students move from repeated addition to skip counting and then to multiplication.

Tell Me a Story

Have students write everyday word problems involving multiplication and division. This helps them understand the actions involved with both operations and the relationship between multiplication and division. It can also help them consolidate the central ideas of multiplication and the relationship between factors (the size of each group and the number of groups) and the product (how many all together). As they become better at creating their own word problems, they can share with a partner or small group of students. This format can also be used for creating division problems.

Multiplication Concentration: 2s, 5s, and 10s

Have students play a concentration game that involves matching multiplication expressions using 2s, 5s, and 10s, the corresponding arrays, and their products. These can be important landmark numbers for solving problems using other factors.

General Recommendations to Support Building Foundations and Initial Strategies for Multiplication and Division

- Use familiar contexts to help students build an understanding of multiplication as a given number of groups with an equal number of objects in each group.

- Begin with small numbers and focus on students' efficiency, flexibility, and accuracy with these numbers.

- Support the transition from thinking additively to thinking multiplicatively by using quick images of equal groups and the array model.

Phase 2: Learning Reasoning Strategies for Multiplication and Division

"What is 9×4?"

"36. I had a head start. I knew $9 + 9$ is 18, so I just doubled that."

There are patterns and strategies in multiplication facts that are dependent upon specific numbers. As a reminder, the strategies described in Section 2 are:

- doubling a factor ("I know $2 \times 8 = 16$, so $4 \times 8 = 32$");

- using derived facts ("I know $5 \times 8 = 40$, so 6×8 is one more group of 8; $6 \times 8 = 48$"); and

- using properties of multiplication (e.g., the distributive property: "I know $8 \times 5 = 40$ and $8 \times 1 = 8$, so $8 \times 6 = 48$").

Vignette: Comparing Strategies

The classroom is at a quiet hum as students engage with partners to think about multiplication combinations they know and combinations they are working on. One pair of students is solving 8×9. They each decide the product is 72. When I ask how they figured out the product, they tell me two different strategies. One student says he knows that 8×8 is 64 and he knows he needs to add another 8. I ask him how he knows he is supposed to add 8 and not 9. He says confidently and with perfect sense-making, "I know I have eight 8s, so I need one more 8 so I will have nine 8s." Another student notes that he first multiplied 8×10 because "that is so easy, 80," and then continues, "I took away 8, so 72." He seems to anticipate the same question I asked his partner and explains, "That way I will have nine 8s—I had an extra 8 when I multiplied by 10." They keep working on other problems, and both seem serious but quietly happy with their solutions.

Number Talks

As with addition and subtraction, the Number Talks protocol for multiplication and division problems requires students to share their strategies and justifications for their answers. Again, you can pose a single problem or a string of problems to emphasize a certain reasoning strategy. Encourage students to understand, apply, and describe to one another their strategies, including using multiplication to solve for a division problem. You can also continue to use images that involve equal groups (as in Figure 3–20).

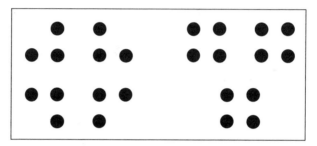

Figure 3–20 Dot images for equal groups

Games

How Many Buttons?

Materials: Two dice, one recording sheet, counters to represent buttons, and six colored sheets of paper to represent shirts. You can also draw or cut the outline of a shirt for each sheet of paper.

Students work with partners.

1. The first die rolled tells the partners how many sheets of paper (shirts) to take.

2. The second die rolled tells them how many "buttons" to put on each shirt.

3. The first partner writes the multiplication equation that matches their representation and shows how many buttons there are altogether. Partners check each other's work and discuss how they solved the equation.

This game reinforces the idea that multiplication is about "How many groups?" and "How many in each group?" As a young teacher, I made a similar game called Bears in Space with cardboard rocket ships and colorful teddy bears. You can also make your own variation!

Multiplication Compare with Factors of 2, 5, and 10

Materials: Two decks of number cards for each pair of students: one with numbers 1–9 and one with numbers 2, 5, and 10.

Students work with partners.

- Each person draws a card from each deck and says the product.
- The partners check each other's answer.
- The person who has the larger product takes the two sets of cards.
- Play five or more rounds.

Variation: Use number dice instead of cards.

Fill It!

Materials: Blank 100 grid (10 × 10), two number dice, and two colored pencils.

This game is played with partners. Two students share a blank 100 grid. The first partner rolls two number dice. The two numbers that appear are the numbers the student uses to make an array on the 100 grids. The first player can draw the array anywhere on the grid but the goal is to fill as many of the 100 spaces as possible. After the player draws the array on the grid, partners identify the multiplication equation that describes the array, and the player records this equation. The second player then rolls the dice, draws a new array, discusses and then records the multiplication equation for this array. The game ends when neither player can make any more arrays. They can then figure out how many empty spaces remain. If playing competitively, the lower score wins.

Strategies for Building Authentic Fact Fluency **85**

You can vary the numbers on the dice. For example, you may want students to first practice simple "times 2" or "times 5" facts and have one die with only the numbers 2 and 5 on it. Another option is to use 12 × 12 grid paper and have one die with the numbers 1–6 and the other with the numbers 5–10. You may also use a deck of cards instead of dice and include a wild card.

Math Journals: Name Your Strategy

As students solve problems that highlight certain strategies, have them write and represent their solutions in a math journal or notebook. For example, given the problem 6 × 4, a student wrote, "I know that 5 × 4 = 20, so I just added one more 4 and got 24." She then noted that she used what she knew about 5s in multiplication to solve the problem. The notebook can be used throughout the year for different math topics, or it can be focused only on multiplication and division fluency.

General Recommendations to Support Learning Reasoning Strategies for Multiplication and Division

- Emphasize that multiplication is about a given number of groups with the same size in each group.
- Continue to link multiplication to skip counting and repeated addition. For some students, this transition takes time.
- Have students connect multiplication expressions to visual representations and to a contextual situation.
- As students describe their strategies, have them justify the strategies with representations as well as expressions and equations. You can also attach a common name to a strategy to help students remember it, such as "breaking apart one factor into smaller factors."
- Provide experiences, problems, and discussions that relate division to multiplication. As students are solving a division problem, look at different student solutions to highlight both division and multiplication equations that match the problem.

Phase 3: Achieving Fluency with Multiplication and Division Facts

On the journey to full fluency with multiplication and division facts, students have moved from understanding foundational concepts related to these operations to learning and applying reasoning strategies. They can quickly solve a division problem by thinking of it as an unknown factor ("I know $6 \times 8 = 48$, so $48 \div 6 = 8$"). They can use known facts to solve less fluent facts (using 2s, 5s, or square numbers). With continued and focused practice, they are ready to "know from memory" all the facts.

Students' self-assessment can help target their practice. As students are engaged in the following activities, ask questions about what they know and what they are working on.

Counting Around the Circle

This activity is from *Number Sense Routines: Building Numerical Literacy Every Day in Grades K–3* (Shumway 2011).

Have students sit in a large circle so that everyone can see each other. Explain the counting sequence (by twos, threes, fives, tens, forward, backward, etc.), and choose a student to start. Going around the circle, each student then says the number that comes next in the sequence.

You may choose to record the sequence as students are stating the numbers. This routine can support students in practicing skip counting both forward and backward on the number line, and in noticing and using number patterns.

Games

Missing Factor

Materials: Deck of cards with equations with missing factors ($6 \times ? = 48$; $7 \times ? = 63$, etc.).

Students turn over a card and say the missing factor. If they can recall it quickly, they put that card in a pile for facts they know.

If they do not know it from memory, they put it in a review pile to practice later. They can use a multiplication table to check their answers.

Students may work with partners.

Multiplication Concentration

Materials: Deck of cards with multiplication expressions on half the cards and their matching products on half the cards.

Place the cards face down in rows and columns. Students take turns, as in the traditional concentration game, finding matching pairs. If students find a match, they state the full equation. They play the game until all the matches are made. Students can also create their own concentration game based on the combinations they are working on.

Multiplication Compare with All Factors

Play Multiplication Compare (see page 86), but have students use two decks of number cards with numbers 1–9.

Multiplication Connect Four

Materials: Multiplication Connect Four game board and two sets of colored chips (e.g., one set of red chips and one set of blue chips).

The game board consists of a five-by-five grid with different multiplication expressions (e.g., 5×7) in each of the twenty-five boxes. Choose expressions that students need practice with or have students write their own multiplication expressions.

Students play in pairs. They choose one expression to place their color chip on for each turn. As they place their chip, they state the given expression and its product. The game continues until one student makes a row or column of four with their color chips.

General Recommendations to Support Achieving Fluency with Multiplication and Division Facts

- Have students practice multiplication and division at the same time to emphasize the relationship between the two operations.
- As with addition and subtraction facts, use familiar games to provide practice and increase speed and efficiency.
- Focus practice on understood but not yet fluent products or unknown factors to speed learning.
- Continue to connect multiplication equations with the area model as appropriate. This will also support students as they move toward two-digit multiplication.

Final Thoughts

As you can see, there are a wealth of ways teachers can help their students be more fluent mathematicians. In addition to the activities and games included in this book, you can find others that match your learning goals. We hope that you will notice a productive buzz in the classroom as your students build upon what they know and share their ideas with others.

When we emphasize foundational concepts and reasoning strategies as the path toward building authentic fluency, students can develop their number sense, articulate their thinking, and understand the reasoning of others. They skip the frenzy of timed tests that contributes to math anxiety and are not afraid of new challenges.

We as teachers and families can listen to students' thinking, acknowledge what students know, and support them in their next steps. We can help diminish anxiety and encourage motivation, engagement, and confidence. Students then become accurate, efficient, and flexible with number combinations and achieve authentic fluency.

AFTERWORD

M. Colleen Cruz

Just recently I was sitting in on a curriculum night, and my ears pricked up as the teacher shifted to talking about math instruction. I heard him say, "We will be doing some timed problem practice that we call 'sprints' in order to work on fluency. I know that it's not the best way to work on fluency. I am aware that there are better ways. But I don't know them yet. So until I learn them, we'll stick with the sprints. But please know that I plan to change this once I know more."

As I sat in that classroom and listened to the teacher speak so frankly with families about engaging kids in a practice he knew wasn't great, I found the teacher refreshingly honest and the situation familiarly frustrating. I wished at the time that I could leap out of my seat and hand this teacher a copy of the book you just finished reading.

But, alas, the copy I had was still a draft and hadn't been published yet.

When Linda Ruiz Davenport, Connie S. Henry, Douglas H. Clements, and Julie Sarama wrote this book, they created a text that directly addresses the feelings this teacher expressed on curriculum night, and those of so many others. When it comes to teaching math fluency to children, there has to be a better way—and these authors have shown us that there is.

Whether we read this book and relive our own childhood struggles over memorization and worksheets, or we read with our current and future students in our hearts, we read knowing that the funny feeling we had about doing math fluency wrong was proven right by research. And better yet, we now have actual real-world strategies for bringing authentic, experiential, and meaningful fluency to our students.

The me that was a third grader doing endless multiplication drills is almost as grateful to Linda, Connie, Douglas, and Julie as the present-day me—the one who is now the parent of a third grader and a kindergartner currently embarking on quests for math fluency. Teachers' knowledge that using fingers and dot images, playing games, and involving families can all work toward the cause of math fluency is going to be revolutionary to so many kids' math lives.

I cannot wait to find that teacher I met at the curriculum night and hand him this book.

APPENDIX

Arithmetic: Mathematical Properties

From the earliest levels, additive arithmetic depends on two properties:

1. The associative law of addition: $(a + b) + c = a + (b + c)$.

    For example, this allows a mental addition strategy that simplifies some computations, such as: $4 + 4 + 6 = 4 + (4 + 6) = 4 + 10 = 14$.

2. The commutative law of addition: $a + b = b + a$.

    Once you know $5 + 3 = 8$, you also know that $3 + 5 = 8$.

Mathematician Roger Howe calls these—taken together—simply the "any which way" laws.

Subtraction does not follow these laws. Subtraction is defined mathematically as the inverse of addition; that is, subtraction is the additive inverse $-a$ for any a, such that $a + - a = 0$. Or, for $8 - 3$, the difference is the number that, when added to 3, results in 8. So, $c - a = b$ means that b is the number that satisfies $a + b = c$. Thus, although it seems cumbersome, one can think of $(8 - 3)$ as $((5 + 3) - 3) = 5 + (3 - 3) = 5 + 0 = 5$. Or, since we know that subtraction and addition are inverses of each other, then saying $8 - 3 = \_\_$ means the same as $8 = 3 + \_\_$.

Multiplication also follows the associative and commutative laws. There is also another law:

3. The distributive law: the product of multiplication is the same when the operation is performed on a whole set as when it is performed on the individual members of the set.

    This can be seen when we express (8×3) as $(5 \times 3) + (3 \times 3) = 15 + 9 = 24$.

Again, division does not follow these laws. But, similarly, division is defined mathematically as the inverse of multiplication. So, $c \div a = b$ means that b is the number that satisfies $a \times b = c$.

REFERENCES

Ashcraft, M. H. 1982. "The Development of Mental Arithmetic: A Chronometric Approach." *Developmental Review* 2: 213–236.

Aubrey, C. 1997. "Children's Early Learning of Number in School and Out." In *Teaching and Learning Early Number*, edited by I. Thompson, 20–29. Philadelphia: Open University Press.

Aunola, K., E. Leskinen, M.-K. Lerkkanen, and J.-E. Nurmi. 2004. "Developmental Dynamics of Math Performance from Pre-school to Grade 2." *Journal of Educational Psychology* 96: 699–713. doi:10.1037/0022-0663.96.4.699.

Barendregt, W., B. Lindström, E. Rietz-Leppänen, I. Holgersson, and T. Ottosson. 2012. "Development and Evaluation of Fingu: A Mathematics iPad Game using Multi-touch Interaction." Paper presented at the Proceedings of the 11th International Conference on Interaction Design and Children, Bremen, Germany.

Baroody, A. J. 1987. "The Development of Counting Strategies for Single-Digit Addition." *Journal for Research in Mathematics Education* 18: 141–157.

———. 2011. "Learning: A Framework." In *Achieving Fluency: Special Education and Mathematics*, edited by F. Fennell, 15–57. Reston, VA: National Council of Teachers of Mathematics.

Baroody, A. J., N. P. Bajwa, and M. Eiland. 2009. "Why Can't Johnny Remember the Basic Facts?" *Developmental Disabilities* 15: 69–79.

Baroody, A. J., and A. Dowker. 2003. *The Development of Arithmetic Concepts and Skills: Constructing Adaptive Expertise*. Mahwah, NJ: Erlbaum.

Baroody, A. J., M. Eiland, and B. Thompson. 2009. "Fostering At-Risk Preschoolers' Number Sense." *Early Education and Development* 20 (1): 80–128. doi:10.1080/10409280802206619.

Baroody, A. J., D. J. Purpura, M. D. Eiland, and E. E. Reid. 2015. "The Impact of Highly and Minimally Guided Discovery Instruction on Promoting the Learning of Reasoning Strategies for Basic Add-1 and Doubles Combinations." *Early Childhood Research Quarterly* 30, part A(0): 93–105. doi:10.1016/j.ecresq.2014.09.003.

Baroody, A. J., and L. Rosu. 2004. *Adaptive Expertise with Basic Addition and Subtraction Combinations—the Number Sense View*. Paper presented at the American Educational Research Association, San Francisco, CA.

Baroody, A. J., and S. H. Tiilikainen. 2003. "Two Perspectives on Addition Development." In *The Development of Arithmetic Concepts and Skills: Constructing Adaptive Expertise*, edited by A. J. Baroody and A. Dowker, 75–125. Mahwah, NJ: Erlbaum.

Berch, D. B., and M. M. M. Mazzocco, eds. 2007. *Why Is Math So Hard for Some Children? The Nature and Origins of Mathematical Learning Difficulties and Disabilities.* Baltimore: Paul H. Brooks.

Bisanz, J., J. A. Sherman, C. Rasmussen, and E. Ho. 2005. "Development of Arithmetic Skills and Knowledge in Preschool Children." In *Handbook of Mathematical Cognition,* edited by J. I. D. Campbell, 143–162. New York: Psychology Press.

Bishop, A. J., and H. J. Forgasz. 2007. "Issues in Access and Equity in Mathematics Education." In *Second Handbook of Research on Mathematics Teaching and Learning,* edited by F. K. Lester Jr., 2:1145–1167. New York: Information Age Publishing.

Boaler, J. 2002. "Paying the Price for 'Sugar and Spice': Shifting the Analytical Lens in Equity Research." *Mathematical Thinking and Learning* 4: 127–144.

———. 2015. "Fluency Without Fear: Research Evidence on the Best Ways to Learn Math Facts." www.youcubed.org/evidence/fluency-without-fear/.

Brownell, W. A. 1928. *The Development of Children's Number Ideas in the Primary Grades.* Chicago: Department of Education, University of Chicago.

Bryant, P. E. 1997. "Mathematical Understanding in the Nursery School Years." In *Learning and Teaching Mathematics: An International Perspective,* edited by T. Nunes and P. Bryant, 53–67. East Sussex, UK: Psychology Press.

Canobi, K. H., R. A. Reeve, and P. E. Pattison. 1998. "The Role of Conceptual Understanding in Children's Addition Problem Solving." *Developmental Psychology* 34: 882–891.

Carpenter, T. P., E. H. Fennema, M. L. Franke, L. Levi, and S. B. Empson. 2014. *Children's Mathematics: Cognitively Guided Instruction.* 2nd ed. Portsmouth, NH: Heinemann.

Carr, M., G. Taasoobshirazi, R. Stroud, and M. Royer. 2011. "Combined Fluency and Cognitive Strategies Instruction Improves Mathematics Achievement in Early Elementary School." *Contemporary Educational Psychology* 36: 323–333.

Celedòn-Pattichis, S., S. A. Peters, L. L. Borden, J. R. Males, S. J. Pape, O. Chapman, D. H. Clements, and J. Leonard. 2018. "Asset-Based Approaches to Equitable Mathematics Education Research and Practice." *Journal for Research in Mathematics Education* 49 (4): 373–389. doi:10.5951/jresematheduc.49.4.0373.

Clements, D. H., and B. K. Nastasi. 1985. "Effects of Computer Environments on Social-Emotional Development: Logo and Computer-Assisted Instruction." *Computers in the Schools* 2 (2–3): 11–31. doi:10.1300/J025v02n02_04.

Clements, D. H., and J. Sarama. 2007. "Effects of a Preschool Mathematics Curriculum: Summative Research on the *Building Blocks* Project." *Journal for Research in Mathematics Education* 38 (2): 136–163.

———. 2007–13. *Building Blocks,* 2 vols. Columbus, OH: McGraw-Hill Education.

———. 2008a. "Experimental Evaluation of the Effects of a Research-Based Preschool Mathematics Curriculum." *American Educational Research Journal* 45 (2): 443–494. doi:10.3102/0002831207312908.

———. 2008b. "Mathematics and Technology: Supporting Learning for Students and Teachers." In *Contemporary Perspectives on Science and Technology in Early Childhood Education*, edited by O. N. Saracho and B. Spodek, 127–147. Charlotte, NC: Information Age.

———. 2014a. *Learning and Teaching Early Math: The Learning Trajectories Approach.* 2nd ed. New York: Routledge.

———. 2014b. "Learning Trajectories: Foundations for Effective, Research-Based Education." In *Learning over Time: Learning Trajectories in Mathematics Education*, edited by A. P. Maloney, J. Confrey, and K. H. Nguyen, 1–30. New York: Information Age Publishing.

Clements, D. H., J. Sarama, and B. L. MacDonald. 2017. "Subitizing: The Neglected Quantifier." In *Constructing Number: Merging Perspectives from Psychology and Mathematics Education*, edited by A. Norton and M. W. Alibali, 13–45. Gateway East, Singapore: Springer.

Clements, D. H., J. Sarama, M. E. Spitler, A. A. Lange, and C. B. Wolfe. 2011. "Mathematics Learned by Young Children in an Intervention Based on Learning Trajectories: A Large-Scale Cluster Randomized Trial." *Journal for Research in Mathematics Education* 42 (2): 127–166. doi:10.5951/jresematheduc.42.2.0127.

Clements, D. H., J. Sarama, C. B. Wolfe, and M. E. Spitler. 2013. "Longitudinal Evaluation of a Scale-Up Model for Teaching Mathematics with Trajectories and Technologies: Persistence of Effects in the Third Year." *American Educational Research Journal* 50 (4): 812–850. doi:10.3102/0002831212469270.

Codding, R. S., A. Hilt-Panahon, C. J. Panahon, and J. L. Benson. 2009. "Addressing Mathematics Computation Problems: A Review of Simple and Moderate Intensity Interventions." *Education and Treatment of Children* 32 (2): 279–312.

Crollen, V., and M.-P. Noël. 2015. "The Role of Fingers in the Development of Counting and Arithmetic Skills." *Acta Psychologica* 156 (0): 37–44. doi:10.1016/j.actpsy.2015.01.007.

Davis, R. B. 1984. *Learning Mathematics: The Cognitive Science Approach to Mathematics Education.* Norwood, NJ: Ablex.

Dowker, A. 2004. *What Works for Children with Mathematical Difficulties?* (Research Report no. 554). Nottingham, UK: University of Oxford, Department for Eduction and Skills.

———. 2005. "Early Identification and Intervention for Students with Mathematics Difficulties." *Journal of Learning Disabilities* 38: 324–332.

Duhon, G. J., S. H. House, and T. A. Stinnett. 2012. "Evaluating the Generalization of Math Fact Fluency Gains Across Paper and Computer Performance Modalities." *Journal of School Psychology* 50: 335–345. doi:10.1016/j.jsp.2012.01.003.

Engel, M., A. Claessens, and M. A. Finch. 2013. "Teaching Students What They Already Know? The (Mis)alignment Between Mathematics Instructional Content and Student Knowledge in Kindergarten." *Educational Evaluation and Policy Analysis* 35 (2): 157–178. doi:10.3102/0162373712461850.

Ericsson, K. A., R. T. Krampe, and C. Tesch-Römer. 1993. "The Role of Deliberate Practice in the Acquisition of Expert Performance." *Psychological Review* 100: 363–406.

Fletcher, J. D., D. E. Hawley, and P. K. Piele. 1990. "Costs, Effects, and Utility of Microcomputer Assisted Instruction in the Classroom." *American Educational Research Journal* 27 (4): 783–806.

Fletcher-Flinn, C. M., and B. Gravatt. 1995. "The Efficacy of Computer Assisted Instruction (CAI): A Meta-analysis." *Journal of Educational Computing Research* 12 (3): 219–242. doi:10.2190/51D4-F6L3-JQHU-9M31.

Fuchs, L. S., D. Fuchs, C. L. Hamlett, S. R. Powell, A. M. Capizzi, and P. M. Seethaler. 2006. "The Effects of Computer-Assisted Instruction on Number Combination Skill in At-Risk First Graders." *Journal of Learning Disabilities* 39: 467–475.

Fuchs, L. S., D. C. Geary, D. L. Compton, D. Fuchs, C. Schatschneider, C. L. Hamlett, J. Deselms, et al. 2013. "Effects of First-Grade Number Knowledge Tutoring with Contrasting Forms of Practice." *Journal of Educational Psychology* 105 (1): 58–77. doi:10.1037/a0030127.

Fuson, K. C. 1992. "Research on Whole Number Addition and Subtraction." In *Handbook of Research on Mathematics Teaching and Learning*, edited by D. A. Grouws, 243–275. New York: Macmillan.

———. 2003. "Developing Mathematical Power in Whole Number Operations." In *A Research Companion to Principles and Standards for School Mathematics*, edited by J. Kilpatrick, W. G. Martin, and D. Schifter, 68–94. Reston, VA: National Council of Teachers of Mathematics.

Fuson, K. C., T. Perry, and Y. Kwon. 1994. "Latino, Anglo, and Korean Children's Finger Addition Methods." In *Research on Learning and Instruction of Mathematics in Kindergarten and Primary School*, edited by J. E. H. Van Luit, 220–228. Doetinchem, Netherlands: Graviant.

Geary, D. C. 2004. "Mathematics and Learning Disabilities." *Journal of Learning Disabilities* 37: 4–15.

Geary, D. C., C. C. Bow-Thomas, and Y. Yao. 1992. "Counting Knowledge and Skill in Cognitive Addition: A Comparison of Normal and Mathematically Disabled Children." *Journal of Experimental Child Psychology* 54: 372–391.

Geary, D. C., S. C. Brown, and V. A. Smaranayake. 1991. "Cognitive Addition: A Short Longitudinal Study of Strategy Choice and Speed-of-Processing Differences in Normal and Mathematically Disabled Children." *Developmental Psychology* 27: 787–797.

Geary, D. C., M. K. Hoard, J. Byrd-Craven, L. Nugent, and C. Numtee. 2007. "Cognitive Mechanisms Underlying Achievement Deficits in Children with Mathematical Learning Disability." *Child Development* 78: 1343–1359.

Geary, D. C., M. K. Hoard, and C. O. Hamson. 1999. "Numerical and Arithmetical Cognition: Patterns of Functions and Deficits in Children at Risk for a Mathematical Disability." *Journal of Experimental Child Psychology* 74: 213–239.

Gersten, R., N. C. Jordan, and J. R. Flojo. 2005. "Early Identification and Interventions for Students with Mathematical Difficulties." *Journal of Learning Disabilities* 38: 293–304.

Gervasoni, A. 2005. "The Diverse Learning Needs of Children Who Were Selected for an Intervention Program." In *Proceedings of the 29th Conference of the International Group for the Psychology of Mathematics Education*, edited by H. L. Chick and J. L. Vincent, 3:33–40. Melbourne, Australia: PME.

Gervasoni, A., T. Hadden, and K. Turkenburg. 2007. "Exploring the Number Knowledge of Children to Inform the Development of a Professional Learning Plan for Teachers in the Ballarat Diocese as a Means of Building Community Capacity." In *Mathematics: Essential Research, Essential Practice* (*Proceedings of the 30th Annual Conference of the Mathematics Education Research Group of Australasia*), edited by J. Watson and K. Beswick, 3:305–314. Hobart, Australia: MERGA.

Gray, E. M., and D. Pitta. 1997. "Number Processing: Qualitative Differences in Thinking and the Role of Imagery." In *Proceedings of the 20th Annual Conference of the Mathematics Education Research Group of Australasia*, edited by L. Puig and A. Gutiérrez, 3:35–42. Rotorua, New Zealand: Mathematics Education Research Group of Australasia.

Groen, G., and L. B. Resnick. 1977. "Can Preschool Children Invent Addition Algorithms?" *Journal of Educational Psychology* 69: 645–652.

Grupe, L. A., and N. W. Bray. 1999. "What Role Do Manipulatives Play in Kindergartners' Accuracy and Strategy Use When Solving Simple Addition Problems?" Paper presented at the Society for Research in Child Development, Albuquerque, NM.

Harskamp, E. 2015. "The Effects of Computer Technology on Primary School Students' Mathematics Achievement: A Meta-analysis." In *The Routledge International Handbook of Dyscalculia*, edited by S. Chinn, 383–392. Abingdon, UK: Routledge.

Hasselbring, T. S., L. I. Goin, and J. D. Bransford. 1988. "Developing Math Automaticity in Learning Handicapped Children: The Role of Computerized Drill and Practice." *Focus on Exceptional Children* 20 (6): 1–7.

Haugland, S. W. 1992. "Effects of Computer Software on Preschool Children's Developmental Gains." *Journal of Computing in Childhood Education* 3 (1): 15–30.

Henry, V. J., and R. S. Brown. 2008. "First-Grade Basic Facts: An Investigation into Teaching and Learning of an Accelerated, High-Demand Memorization Standard." *Journal for Research in Mathematics Education* 39 (2): 153–183.

Hiebert, J. C., and D. A. Grouws. 2007. "The Effects of Classroom Mathematics Teaching on Students' Learning." In *Second Handbook of Research on Mathematics Teaching and Learning*, edited by F. K. Lester Jr., 1:371–404. New York,: Information Age Publishing.

Holmes, J., S. E. Gathercole, and D. L. Dunning. 2009. "Adaptive Training Leads to Sustained Enhancement of Poor Working Memory in Children." *Developmental Science* 12 (4): 9–15.

Hopkins, S. L., and M. J. Lawson. 2004. "Explaining Variability in Retrieval Times for Addition Produced by Students with Mathematical Learning Difficulties." In *Proceedings of the 28th Conference of the International Group for the Psychology in Mathematics Education*, edited by M. J. Høines and A. B. Fuglestad, 3:57–64. Bergen, Norway: Bergen University College.

Hunting, R., and G. Davis, eds. 1991. *Early Fraction Learning*. New York: Springer-Verlag.

Illustrative Mathematics. n.d.a. "Hitting the Target Number." Accessed November 1, 2018. www.illustrativemathematics.org/content-standards/2/OA/B/2/tasks/1396.

———. n.d.b. "Valid Inequalities?" Accessed November 1, 2018. www.illustrativemathematics.org/content-standards/tasks/466.

Kamii, C., and A. Dominick. 1998. "The Harmful Effects of Algorithms in Grades 1–4." In *The Teaching and Learning of Algorithms in School Mathematics*, edited by L. J. Morrow and M. J. Kenney, 130–140. Reston, VA: National Council of Teachers of Mathematics.

Ketamo, H., and K. Kiili. 2010. "Conceptual Change Takes Time: Game Based Learning Cannot Be Only Supplementary Amusement." *Journal of Educational Multimedia and Hypermedia* 19 (4): 399–419.

Kraus, W. H. 1981. "Using a Computer Game to Reinforce Skills in Addition Basic Facts in Second Grade." *Journal for Research in Mathematics Education* 12: 152–155.

Landerl, K., A. Bevan, and B. Butterworth. 2004. "Developmental Dyscalculia and Basic Numerical Capacities: A Study of 8–9-year-old children." *Cognition* 93: 99–125.

Lavin, R. J., and J. E. Sanders. 1983. "Longitudinal Evaluation of the C/A/I Computer Assisted Instruction Title 1 Project: 1979–82." Retrieved from Chelmsford, MA: Merrimack Education Center website.

LeFevre, J.-A., G. S. Sadeskey, and J. Bisanz. 1996. "Selection of Procedures in Mental Addition: Reassessing the Problem Size Effect in Adults." *Journal of Experimental Psychology: Learning, Memory, and Cognition* 22: 216–230.

Lerner, J. 1997. *Learning Disabilities*. Boston: Houghton Mifflin.

Lesh, R., T. Post, and M. Behr. 1987. "Representations and Translations Among

Representations in Mathematics Learning and Problem Solving." *Problems of Representation in the Teaching and Learning of Mathematics* 21: 33–40.

Mazzocco, M. M. M., and G. F. Myers. 2003. "Complexities in Identifying and Defining Mathematics Learning Disability in the Primary School-Age Years." *Annals of Dyslexia* 53: 218–253.

McNeil, N. M. 2008. "Limitations to Teaching Children 2 + 2 = 4: Typical Arithmetic Problems Can Hinder Learning of Mathematical Equivalence." *Child Development* 79 (5): 1524–1537.

McNeil, N. M., D. L. Chesney, P. G. Matthews, E. R. Fyfe, L. A. Petersen, A. E. Dunwiddie, and M. C. Wheeler. 2012. "It Pays to Be Organized: Organizing Arithmetic Practice Around Equivalent Values Facilitates Understanding of Math Equivalence." *Journal of Educational Psychology* 104 (4): 1109–1121. doi:10.1037/a0028997.

McNeil, N. M., E. R. Fyfe, and A. E. Dunwiddie. 2015. "Arithmetic Practice Can Be Modified to Promote Understanding of Mathematical Equivalence." *Journal of Educational Psychology* 107 (2): 423–436. doi:10.1037/a0037687.

McNeil, N. M., E. R. Fyfe, L. A. Petersen, A. E. Dunwiddie, and H. Brletic-Shipley. 2011. "Benefits of Practicing 4 = 2 + 2: Nontraditional Problem Formats Facilitate Children's Understanding of Mathematical Equivalence." *Child Development* 82 (5): 1620–1633.

Miller, K. F. 1984. "Child as the Measurer of All Things: Measurement Procedures and the Development of Quantitative Concepts." In *Origins of Cognitive Skills: The Eighteenth Annual Carnegie Symposium on Cognition*, edited by C. Sophian, 193–228. Hillsdale, NJ: Erlbaum.

Moll, L. C., C. Amanti, D. Neff, and N. Gonzalez. 1992. "Funds of Knowledge for Teaching: Using a Qualitative Approach to Connect Homes and Classrooms." *Theory into Practice* 31: 132–141.

Murata, A. 2004. "Paths to Learning Ten-Structured Understanding of Teen Sums: Addition Solution Methods of Japanese Grade 1 Students." *Cognition and Instruction* 22: 185–218.

Murata, A., and K. C. Fuson. 2006. "Teaching as Assisting Individual Constructive Paths Within an Interdependent Class Learning Zone: Japanese First Graders Learning to Add Using 10." *Journal for Research in Mathematics Education* 37 (5): 421–456. doi:10.2307/30034861.

National Council of Teachers of Mathematics. 2014. *Principles to Actions: Ensuring Mathematical Success for All*. Reston, VA: National Council of Teachers of Mathematics.

———. n.d. "Product Game." *NCTM Illuminations*. www.nctm.org/Classroom-Resources /Illuminations/Interactives/Product-Game/.

National Mathematics Advisory Panel. 2008. *Foundations for Success: The Final Report of the National Mathematics Advisory Panel.* Washington, DC: U.S. Department of Education, Office of Planning, Evaluation, and Policy Development.

National Research Council. 2001. *Adding It Up: Helping Children Learn Mathematics.* Washington, DC: National Academy Press. doi:10.17226/9822.

Niemiec, R. P., and H. J. Walberg. 1987. "Comparative Effects of Computer-Assisted Instruction: A Synthesis of Reviews." *Journal of Educational Computing Research* 3 (1): 19–37. doi:: https://doi.org/10.2190/RMX5-1LTB-QDCC-D5HA.

Ostad, S. A. 1998. "Subtraction Strategies in Developmental Perspective: A Comparison of Mathematically Normal and Mathematically Disabled Children." In *Proceedings of the 22nd Conference for the International Group for the Psychology of Mathematics Education,* edited by A. Olivier and K. Newstead, 3: 311–318. Stellenbosch, South Africa: University of Stellenbosch.

Pearson. 2017. *Investigations in Number, Data, and Space.* 3rd ed. (*Investigations 3.*) Boston: Pearson.

Poincare, H. (1905). *Science and Hypothesis.* New York, NY: Walter Scott.

Powell, S. R., L. S. Fuchs, and D. Fuchs. 2011. "Number Combinations Remediation for Students with Mathematics Difficulty." *Perspectives on Language and Literacy* 37 (2): 11–16.

Price, A. J. 2001. "Atomistic and Holistic Approaches to the Early Primary Mathematics Curriculum for Addition." In *Proceedings of the 25th Conference of the International Group for the Psychology in Mathematics Education,* edited by M. Van den Heuvel-Panhuizen, 4: 73–80. Utrecht, Netherlands: Freudenthal Institute.

Primavera, J., P. P. Wiederlight, and T. M. DiGiacomo. 2001. *Technology Access for Low-Income Preschoolers: Bridging the Digital Divide.* Paper presented at the American Psychological Association, San Francisco, CA.

Ragosta, M., P. Holland, and D. T. Jamison. 1981. *Computer-Assisted Instruction and Compensatory Education: The ETS/LAUSD Study.* Princeton, NJ: Educational Testing Service.

Rittle-Johnson, B., and M. W. Alibali. 1999. "Conceptual and Procedural Knowledge of Mathematics: Does One Lead to the Other?" *Journal of Educational Psychology* 91: 175–189.

Sarama, J. 2002. "Listening to Teachers: Planning for Professional Development." *Teaching Children Mathematics* 9 (1): 36–39.

Sarama, J., and D. H. Clements. 2009. *Early Childhood Mathematics Education Research: Learning Trajectories for Young Children.* New York: Routledge.

————. Forthcoming. "Promoting a Good Start: Technology in Early Childhood Mathematics." In *Promising Models to Improve Primary Mathematics Learning in Latin America and the Caribbean Using Technology*, edited by E. Arias, J. Cristia, and S. Cueto. Washington, DC: Inter-American Development Bank.

Sarama, J., and A.-M. DiBiase. 2004. "The Professional Development Challenge in Preschool Mathematics." In *Engaging Young Children in Mathematics: Standards for Early Childhood Mathematics Education*, edited by D. H. Clements, J. Sarama, and A.-M. DiBiase, 415–446. Mahwah, NJ: Erlbaum.

Shih, J., W. R. Speer, and B. C. Babbitt. 2011. "Instruction: Yesterday I Learned to Add; Today I Forgot." In *Achieving Fluency: Special Education and Mathematics*, edited by F. Fennell, 59–83. Reston, VA: National Council of Teachers of Mathematics.

Shin, N., L. M. Sutherland, C. A. Norris, and E. Soloway. 2012. "Effects of Game Technology on Elementary Student Learning in Mathematics." *British Journal of Educational Technology* 43 (4): 540–560. doi:10.1111/j.1467-8535.2011.01197.x.

Shumway, J. F. 2011. *Number Sense Routines: Building Numerical Literacy Every Day in Grades K–3*. Portland, ME: Stenhouse.

Siegler, R. S. 1993. "Adaptive and Non-adaptive Characteristics of Low Income Children's Strategy Use." In *Contributions of Psychology to Science and Mathematics Education*, edited by L. A. Penner, G. M. Batsche, H. M. Knoff, and D. L. Nelson, 341–366. Washington, DC: American Psychological Association.

————. 1995. "How Does Change Occur: A Microgenetic Study of Number Conservation." *Cognitive Psychology* 28: 255–273. doi:10.1006/cogp.1995.1006.

Smith, C. R., N. E. Marchand-Martella, and R. C. Martella. 2011. "Assessing the Effects of the *Rocket Math* Program with a Primary Elementary School Student at Risk for School Failure: A Case Study." *Education and Treatment of Children* 34: 247–258.

Smith, M. S., and M. K. Stein. 2011. *5 Practices for Orchestrating Productive Mathematics Discussions.* Reston, VA: National Council of Teachers of Mathematics.

Steffe, L. P. 1994. "Children's Multiplying Schemes." In *The Development of Multiplicative Reasoning in the Learning of Mathematics*, edited by G. Harel and J. Confrey, 3–39. Albany, NY: SUNY Press.

Steffe, L. P., and P. Cobb. 1988. *Construction of Arithmetical Meanings and Strategies.* New York: Springer-Verlag.

Van den Heuvel-Panhuizen, M. 1990. "Realistic Arithmetic/Mathematics Instruction and Tests." In *Contexts Free Productions Tests and Geometry in Realistic Mathematics Education*, edited by K. P. E. Gravemeijer, M. Van den Heuvel-Panhuizen, and L. Streefland, 53–78. Utrecht, Netherlands: OWandOC.

More titles in the
Not This, But That Series

No More Culturally Irrelevant Teaching

Mariana Souto-Manning, Carmen Lugo Llerena, Jessica Martell, Abigail Salas Maguire, and Alicia Arce-Boardman

Grades K–5 / 978-0-325-08979-9 / 96pp

No More Mindless Homework

Kathy Collins and Janine Bempechat

Grades K–5 / 978-0-325-09281-2 / 96pp

No More Independent Reading Without Support

Debbie Miller and Barbara Moss

Grades K–6 / 978-0-325-04904-5 / 96pp

DEDICATED TO TEACHERS

Heinemann.com

  @HeinemannPub